CONTROL VALVE PRIMER

A User's Guide

Hans D. Baumann

Instrument Society of America

CONTROL VALVE PRIMER
A User's Guide

Printed in the United States of America

In preparing this work the author and publishers have not investigated or considered patents which may apply to the subject matter hereof. It is the responsibility of the readers and users of the subject matter to protect themselves against liability for infringement of patents. The information and recommendations contained herein are not intended for specific applications but are of a general educational nature. Accordingly, the authors and publishers assume no responsibility and disclaim all liability of any kind, however arising, as a result of using the subject matter of this work.

The equipment referenced in this work has been selected by the author as examples of the technology; no endorsement of any product is intended by the publisher. In all instances, the manufacturer's procedures should prevail regarding the use of specific equipment. No representation, expressed or implied, is made with regard to the availability of any equipment, process, formula or other procedures contained herein.

While the statements and comments in each chapter are based on the author's experience and views, none are to be construed as recommendations of one product over another.

INSTRUMENT SOCIETY OF AMERICA
67 Alexander Drive
P.O. Box 12277
Research Triangle Park, NC 27709

Library of Congress Cataloging-in-Publication Data

Baumann, Hans D., 1930–
Control valve primer: a user's guide / by Hans D. Baumann.
p. cm.
Includes bibliographical references.
ISBN 1-55617-323-7
1. Pneumatic control valves—Handbooks, manuals, etc. I. Title.
TJ223.V3B38 1991 91-28927
629.8'045—dc20 CIP

DEDICATION

This book is dedicated to Donald L. McDowell, P.E., for providing motivation and to Catherine E. Brashear for her tireless work on the word processor. My thanks also to Gregory McMillan and Stanley Weiner for their valuable critique.

CONTENTS

FOREWORD

This book is intended primarily for those novice instrument engineers who, even though they have been crammed full of feedback theory from their college control courses and consider themselves experts in computer programing, are suddenly faced with the challenge of designing a control loop in a process plant. Here, they discover that, in addition to all the vaguely familiar electronic hardware such as controllers and transmitters, they also need a final control element. This turns out to be a cross between a piece of plumbing hardware and a pneumatic logic system. Such people are usually ill-equipped to answer the 30 plus entry questions on a typical control valve specification sheet in the allotted time of, say, ten minutes. Therefore, having a reference book at their disposal might make their jobs easier, and it might even help them to distinguish what is relevant when bombarded with conflicting claims from competing control valve vendors. At the very least, it may keep them from wasting valuable time by not having to wade through many hundreds of pages of scientifically worded handbook trivia.

While it is true that any personal computer can handle complex sizing equations, I still feel it is useful to have some simple formulas that can be handled by a pocket calculator if the need arises. This led me to propose short cuts in the sizing and noise equations.

The reader is hereby cautioned that the opinions expressed herein are strictly those of the writer, and no guarantee is given, nor implied, as to the accuracy of the book's contents. As a matter of fact, some of the statements are considered downright controversial. I am seldom accused of being unbiased.

It all started when, as president of one of the smaller control valve companies, I visited a major tobacco company in Richmond, Virginia, in order to do some sales promotion. However, after having carried about half a dozen valve samples up to the second floor conference room, the leader of the group of young instrument engineers stated flatly, "Please, no sales talk. Instead, answer some of my questions." As it turned out, he liked the answers and at the end of the session suggested I write a book. Well, here it is.

ABOUT THE AUTHOR

Hans D. Baumann received an industrial engineering education in his native Germany and further studied under U.S. Government sponsorship at Western Reserve University and later at Northeastern University, culminating in a Ph.D. in Mechanical Engineering from Columbia Pacific University. In addition, he is registered as a Professional Engineer in four states. During his professional career, he personally designed or directed the development of over 28 valve lines (see artwork on cover). One of them, the famous "CAMFLEX" valve and its derivations, is produced in eight countries where over two million units have been sold. He is credited with over 114 U.S. and worldwide patents and has published 68 papers and articles in addition to co-authoring four handbooks on valves, instrumentation, and noise.

Prior to founding his own control valve manufacturing company in 1977, Dr. Baumann was an International Consultant, Corporate Vice-President of Masoneilan-International, Inc., Manager of R&D at Worthington S/A in France, Director of Engineering at CASHCO, Inc., and Chief Engineer at W. & T. Co., a valve supplier to the Siemens Co. in Germany.

Usually ahead of his time, his "critical flow factor" and "pipe reducer correction factors" (F_L & F_p), for valve sizing, introduced in the early sixties, later became part of the ISA sizing standards in 1972; his proposal in 1970 to utilize modified jet noise theories for aerodynamic valve noise prediction became the basis for ISA-S75.17-1989. His involvement with standards include serving as Director of the ISA Standards & Practices Dept., Chairman of the ISA SP75.11 Committee, and "U.S. Technical Expert" for IEC Committee SC65B/WG9, in addition to being the Standards Chairman for Control Valves for the Fluid Controls Institute, and a member of the ASME Bioprocessing Equipment Committee. As a well received guest speaker in the U.S. and abroad, he has also been invited as a guest Professor to Kobe University in Japan and to the Korean Advanced Institute of Technology in Korea.

As a Fellow Member of ISA, he has been honored by the Instrument Society of America for his "many contributions to the science and technology of control valves" with the "Chet Beard" and the "UOP Technology" awards; he was named Honorary Member of the Spanish Chemical Engrg. Society; and his many valve designs have been honored with a gold medal from Germany, prizes from France and Japan, and seven U.S. "Vaaler" awards.

Dr. Baumann, P.E., is currently the President of H.D. Baumann, Assoc., Ltd., of Portsmouth, NH.

1

WHAT IS A CONTROL VALVE AND HOW DOES IT AFFECT MY CONTROL LOOP?

Control valves may be the most important, but sometimes the most neglected, part of a control loop. The reason is usually the instrument engineer's unfamiliarity with the many facets, terminologies, and areas of engineering disciplines such as fluid mechanics, metallurgy, noise control, and piping and vessel design that can be involved depending on the severity of service conditions.

Any control loop usually consists of a sensor of the process condition, a transmitter, and a controller that compares the "process variable" received from the transmitter with the "set point," i.e., the desired process condition. The controller, in turn, sends a corrective signal to the "final control element," the last part of the loop and the "muscle" of the process control system.

Control valves are the most common type of final control element; however, there are other types, such as:

- devices that regulate (throttle) electric energy such as silicon-controlled rectifiers,
- variable speed drives,
- feeders, pumps, and belt drives, and
- dampers.

Some of these devices perform functions similar to control valves and could be used as an alternative. For example, in order to control the pH level, a variable stroke-type metering pump may be used to inject acid into wastewater (instead of a control valve lined with polytetrafluorethylene (PTFE)).

What then is a control valve? This is a difficult question since there is considerable overlap with other types of valves. For example, a valve operating strictly in the on-off mode (such as the hydronic solenoid valve in your home heating system) could be replaced by a simple ball valve operated by a pneumatic cylinder, a type usually referred to as an "automated valve."

The distinction between "automated" and "control" valves is usually considered to be the ability of the latter to "modulate," i.e., to assume an infinite number of "throttling" travel positions during normal control service.

Physically, there are three basic components of a control valve:

- The valve body subassembly. This is the working part and, in itself, a pressure vessel.
- The actuator. The device that positions the throttling element inside the valve body.
- Accessories. These are positioners, I/P transducers, limit switches, handwheels, air sets, position sensors, solenoid valves, and travel stops.

A more detailed breakdown of the various types of valves, actuators, and positioners is shown in Figure 1-1.

Now let's discuss what the control valve should do. Referring to Figure 1-2, which shows a somewhat simplified process control loop diagram (omitting the transmitter portion—the controller does the sensing), we see three important function blocks above the control valve symbol.

The first one is control valve gain, K_v. This is determined by the "installed flow characteristic" of the valve (quite different from the characteristic shown in the vendor's catalog). K_v tells you how much the flow through the valve is changing per a given signal change. For input, see Chapter 7.

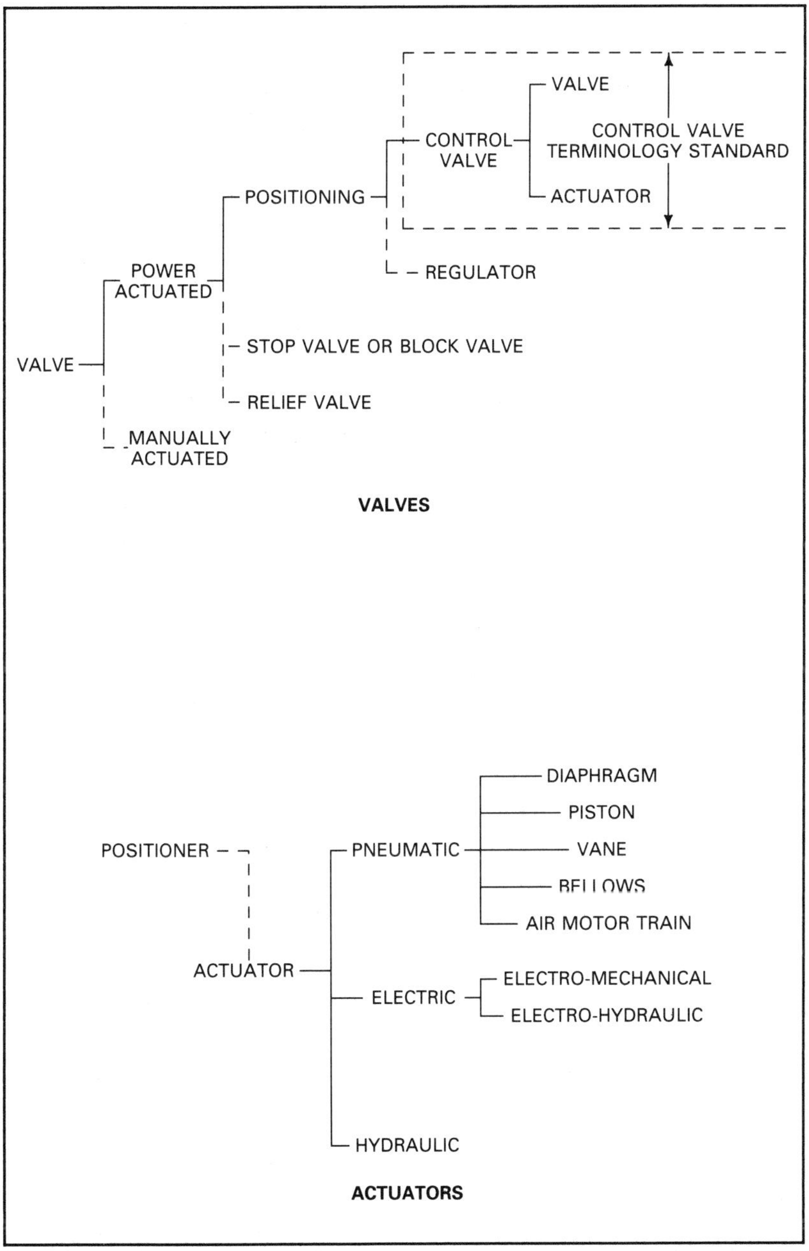

Figure 1-1. Basic control valve terminology from ISA ANSI/ISA-S75.05-1986, by permission of the Instrument Society of America.

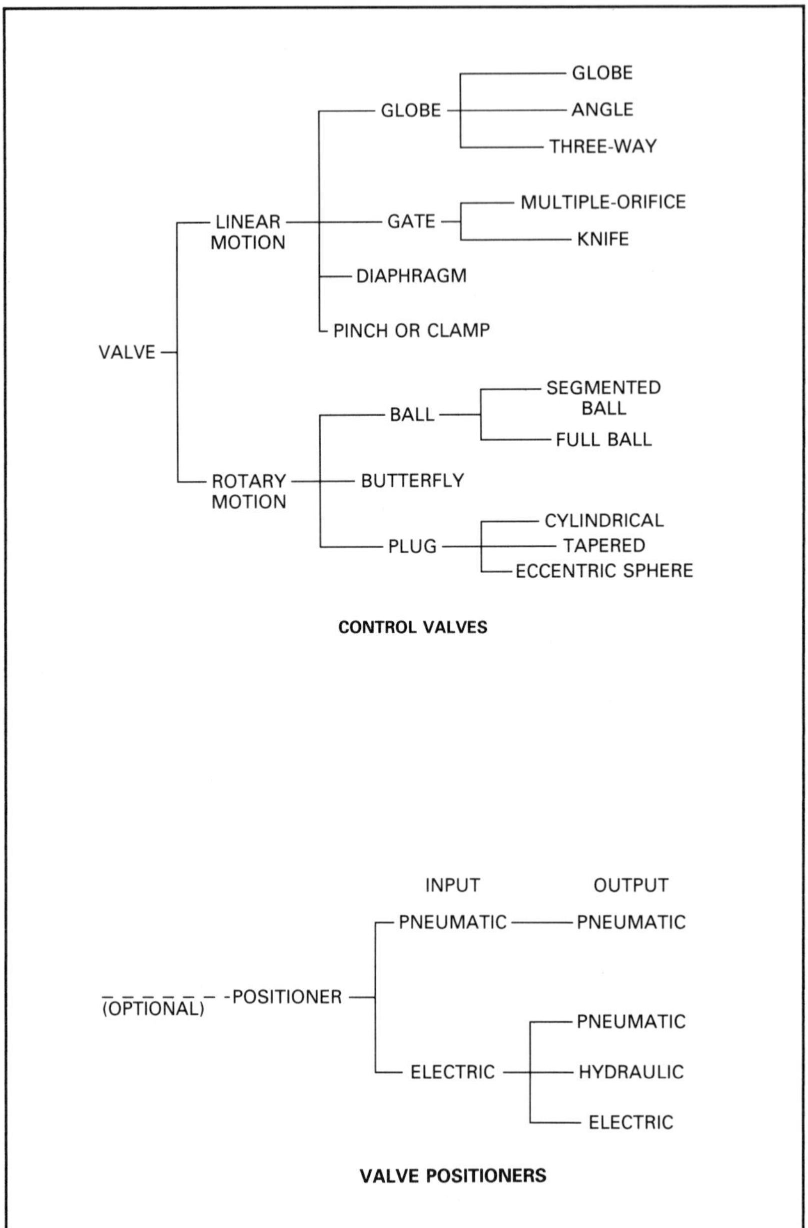

Figure 1-1. *(Continued)*

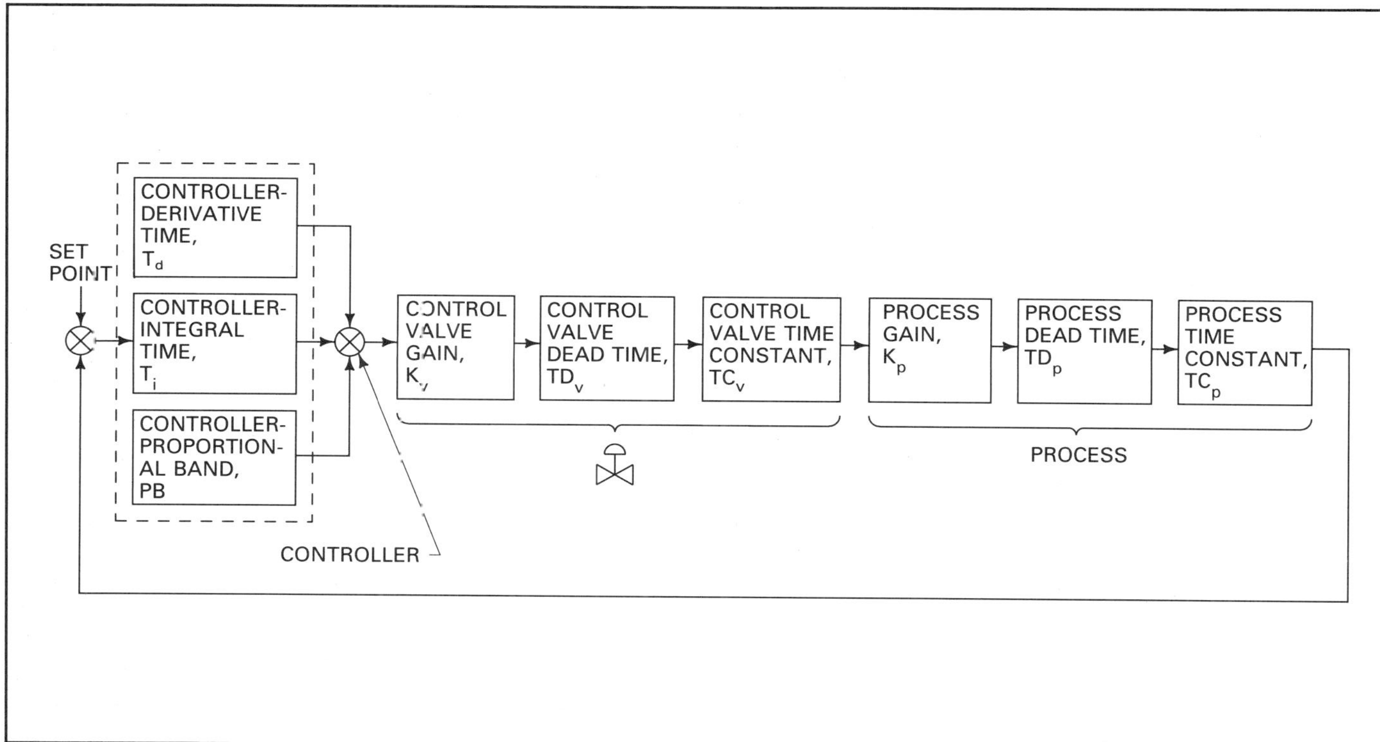

Figure 1-2. Schematic block diagram of controller, control valve, and process in a control loop.

The second block shows the control valve dead time, TD_v. This is the time it takes before a valve moves following a controller signal change. This is usually determined by the valve and actuator friction but may include time lags due to long pneumatic signal transmission lines and the time to build the pressure up in a diaphragm case, for example.

Finally, the third block shows the time constant of the valve, TC_v. This is simply related to the stroking speed of the actuator or actuator/positioner combination (see Chapter 8); i.e., how fast the valve is responding to an upset in your controlled variable. All these function blocks interact, and each one should be considered in evaluating a control valve application.

The "ideal" valve should have a constant gain throughout the flow range, i.e., a linear "installed" flow characteristic, no dead time with packing tightened, and a time constant that is different from that of the process by at least a factor of three.

Need I tell you that there is no such thing as the "ideal" control valve? So let's attempt to develop a workable compromise.

WHAT TO LOOK FOR IN A GOOD CONTROL VALVE DESIGN

Besides the obvious, such as good quality workmanship, correct selection of materials, etc., special attention should be paid to two areas:

- Low dead band of the actuator/valve combination (with tight packing).
- Tight shutoff, in cases of single-seated globe valves and some rotary valves (if required).

The prime concern of an operator of a process control loop is to have a loop that is stable. (Nothing makes people more nervous than a lot of red ink and scattered lines on a strip of paper from a recorder.) The final control element will influence the stability of a loop more than all the other control elements combined.

The biggest culprit here is "dead time."[1] This is the time it takes for the controller to vary the output signal sufficiently to make the actuator and the valve move to a new position. What we are talking about here is that the dead time, TD_v, is essentially the "dead band"* of the actuator/valve combination, or, in case a positioner is used, the dead band of the valve divided by the "open loop gain"** of the positioner plus the positioner's dead band (dead band keeps the valve from responding instantly when the signal changes, which, in turn, causes dead time). The valve itself should never have a dead band of more than 5% of span, that is, less than 0.6 psi for a 3-15 psi signal span. The positioner/valve combination should have no more than 0.5% of signal span. Ignoring process dynamics, a positioner may therefore improve matters by an order of magnitude. However, positioners can raise havoc with the dynamics of a control loop (see Chapter 8 on positioners). The ideal valve is still the one in which the operating dead band with tight stem packing is less than 1%. There you have a maximum stability without the extra cost and complication of having to use a valve positioner.

The very first control valves recognized the value of low dead band. The valve in Figure 1-3, for example, features ball bearings around the actuator stem in order to reduce friction (circa 1932). Low seat leakage can be very beneficial. First, it saves the extra expense of a shutoff valve in case the system closes down. (Note: For safety reasons, never rely on the control valve for an absolute, tight shutoff.) Second, it saves energy. Third, it is an unqualified requirement in temperature control applications in a batch process, especially when you are supplying a heating fluid to a chemical that can undergo an exothermic reaction.

Another feature is good rangeability. As you probably already know, valves are usually terribly oversized. That is, they operate only at perhaps 30% of their rated C_v under normal flow conditions. Furthermore, it is not wise to operate a conventional valve trim at less than 5% travel.† The reason is that the controller might be slightly

---

*This is often confused with "hysteresis," a dead band produced by moving the actuator up and down through 100% travel. Hysteresis is not very important in a "closed-loop" system.

**Usually 50 to 200. Consult vendor's published positioner catalog.

†Without a positioner and with high packing friction, such as graphite packing, the minimum travel position should be higher than the dead band caused by the packing friction (usually 7 to 10%)!

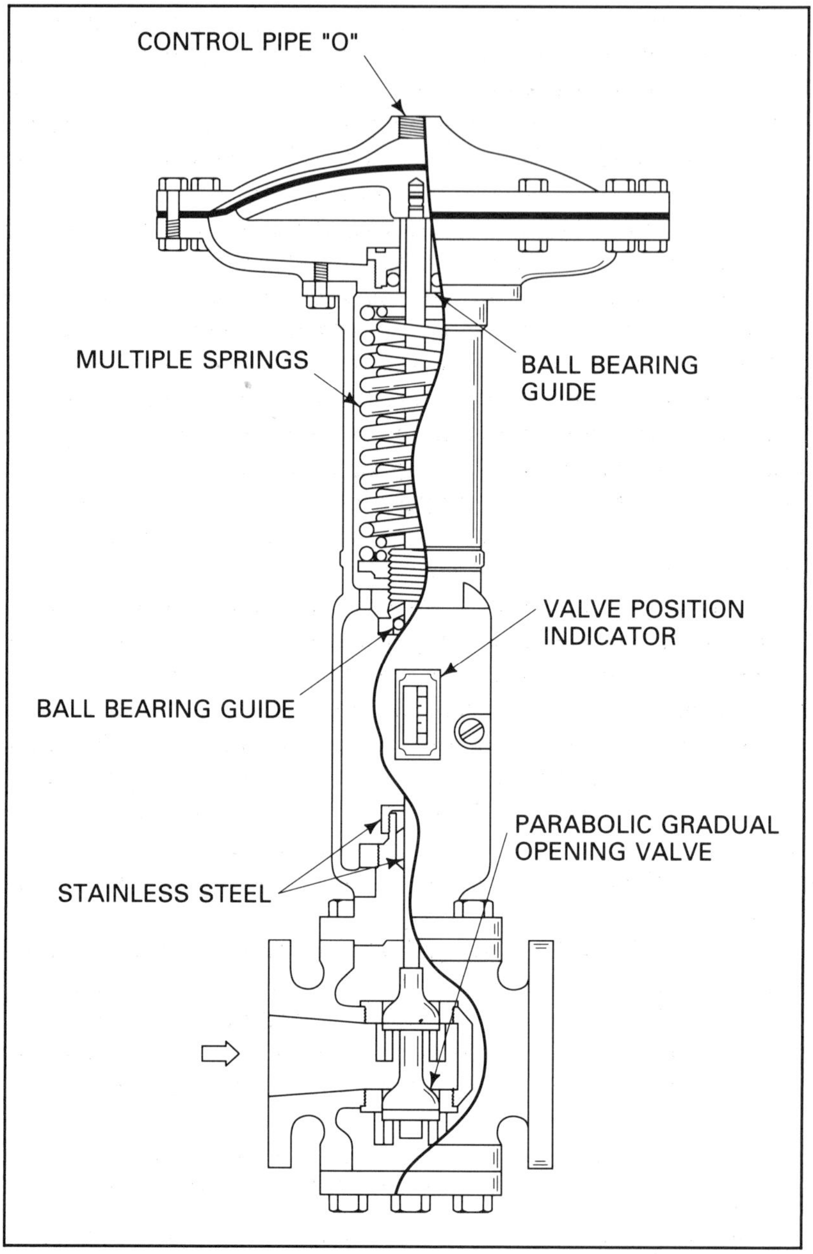

Figure 1-3. Cross section of an early spring-diaphragm actuated control valve, circa 1932, made by the Neilan Co. of Los Angeles, CA. Actuator signal was 2-15 psi.

unstable at the low flow rates, and the actuator, following the sinusoidal output of the controller signal, will push the plug against the seat. When this happens, the actuator and positioner will bleed all the air out and you end up having a really big dead time (before the valve gets moving) again upon signal reversal. In short, you have a big mess and no control.

Using the 5% travel limit, we find that the C_v is only about 7% of the rated C_v with a linear characteristic and not much less with most equal percentage characterized plugs. This gives a useful range for the above normal C_v of 30% divided by 7%, which is 4.5:1. However, in most loops the pressure drop across the valve always varies inversely with the flow, that is, a low ΔP at maximum flow and a high ΔP with low flow. We, therefore, might find that the actual controllable ratio of maximum to minimum flow is perhaps only 3:1. Incidentally, this ratio is the so-called "installed" rangeability, which as you can see is quite different from the "inherent" rangeability shown in a manufacturer's catalog for the same valve. A simple way to remember is that the "inherent" valve rangeability is the ratio of maximum to minimum controllable C_v, while the "installed" rangeability is the ratio of maximum to minimum controllable flow rate in your loop.

Reference

1. McMillan, G. K., and Weiner, S., *How To Become an Instrument Engineer—The Making of a Primadonna*, Research Triangle Park, NC: Instrument Society of America (1987).

2

WHY NOT USE A SPEED-CONTROLLED PUMP?

When the means to electronically control the speed of an electric motor became available in the mid-70s, a thought immediately occurred: "Why not vary the speed or number of revolutions of pumps in order to control the flow rate of liquids in process control systems?" As a matter of fact, this seemingly ideal solution led to the purchase of a large electronics company by the owners of a major oil company.

Alas, things did not turn out to be that simple. For one thing, modern sizing of control valves, coupled with the introduction of rotary control valves having higher inherent flow capacities, led to the use of pumps with smaller head pressure, hence, less "waste" of energy, i.e., pressure drop across the control valve. Energy seems "wasted" to us because the throttling action of a valve is a thermodynamically nonreversible process and, therefore, increases the "entropy" of the fluid; thus, the energy within the fluid is at a lower, less available state.

Unfortunately, as it turned out, reducing the speed of electric motors was not as economical as first thought. Following a flow reduction, most of the energy savings at the pump is converted into useless heat within the motor, variable frequency, and/or variable voltage generators. Economic arguments for speed-controlled electric motor-driven pumps can be made only in systems where the pump is deliberately oversized,[1] forcing the control valve to reduce perhaps 30% of the available pump head pressure at the rated flow rate. This is simply bad control valve sizing. It should be only 5% to 10% (see Chapter 5 on Sizing). Keep in mind that in certain pipe lines the control valve pressure drop is as low as $\frac{1}{2}$% of the total system pressure! An advantage for speed-controlled pumps exists where the static pressure of the system is relatively low. However, this is a relatively rare occurrence.

Other arguments against electric speed control of pumps are:[2]

- high capital investment (averages twice the price of a valve),
- poor dynamic performance compared to small- and medium-sized valves, and
- higher maintenance requirements.

While I might seem to be very negative on this subject, there are indeed applications in which a speed-controlled pump or fan is preferred. For example, if you pump a highly corrosive fluid (and where process dynamics allow), the expense of a special alloy valve can be avoided. Other cases can be made for dangerous fluids such as molten metal, etc. Speed control can improve controller performance due to the absence of valve dead band on applications such as furnace pressure control.

Sometimes the combination of pump speed adjustment (by mechanical means, for example) and a control valve can be beneficial in order to match the pump head to ideal process requirements.

Quite another argument can be made if the pump speed control is through energy sources other than relatively expensive electricity. A case can therefore be made for the diesel engine-driven pumps used to control pressure in oil pipe lines or for utilizing steam in a power plant to feed a turbine-driven feed water pump replacing the former control valve.

References

1. Baumann, H. D., ''Control Valve vs. Variable-Speed Pump,'' *Chemical Engineering*, pp 81–84, June 29 (1981).
2. Wood, W. R., ''Beware of Pitfalls When Applying Variable-Frequency Drives,'' *Power*, pp 47–49, February (1987).

3

WHAT VALVE TYPE SHALL I CHOOSE?

SELECTING THE RIGHT VALVE TYPE

While performance is certainly number one in our selection criteria, cost cannot be ignored since we all must live within our budgets. After having reviewed a number of bids/proposals, you will soon discover that rotary-type control valves have a definite cost advantage in sizes above 2 inches (50 mm), excluding high pressure and special-purpose valves. They tend to have less packing leakage if designed properly (see Chapter 12). This then narrows our focus towards globe valves for general service applications in sizes $\frac{1}{2}$ in. (12.5 mm) to 2 in. (50 mm) and rotary valves in sizes above 2 inches (50 mm). If you select a globe valve, try to specify one with a bolted-on bonnet to facilitate maintenance. Bronze or alloy valves may have threaded bonnets.

In order not to box yourself in price-wise, try not to specify what valve type should be quoted, but rather specify what features *do not work* in your given application. For example, if you have to control crude oil, specify that "no cage-guided trim" is allowed (The solid particles may jam the plug guide!). If you handle hazardous fluids, specify "no threaded end connections," and so on. Specify tight shutoff only if really necessary; it may cost extra money.

Only you know the corrosive nature of your process fluid best; therefore, specify stainless steel, alloy, or plastic as body or trim material. However, for installations in process plants, you may want to refrain from using all-plastic valve bodies. Remember, anything that can be used by a maintenance person as a step ladder will be used as such!

When using a plastic-lined metal valve body, make sure the surrounding housing is pressure-tight and that a secondary stem packing and telltale connection are provided to protect yourself against liner or plastic bellows failure. Bellows made of tetrafluoroethylene (TFE)—Teflon™ plastic—have particularly limited life cycles and are permeable to chlorine.

Valve end connections may be threaded in sizes 2 in. and below for nonhazardous fluids but should be flanged for the handling of chemicals. High pressure steam valves are usually welded, as are some aseptic valves in bioprocessing systems.

While it is usually "OK" to have your favorite valve vendor select the valve type (which saves my adding all the usual catalog pages to this volume), base your final judgment on the criteria mentioned herein.

The maintenance and past experience of a given type of valve is always an important consideration. One of the most embarrassing things that can happen is seeing a valve you chose six months earlier on the trash heap and then listening to unflattering remarks by the maintenance foreman.

By far, the most often encountered causes of premature valve failure are:

- cavitation damage,
- wire drawing of trim,
- erosion of trim,
- resonating plug vibration,
- corrosion, and
- excessive stem packing leakage.

The first can be avoided by selecting a low pressure recovery trim (high F_L), by installing a secondary flow restrictor, or by choosing a better valve location. The second is usually encountered with wet steam (condensate actually). Here, the carbon steel of a valve housing gets eaten away underneath the stainless steel seat ring if there is no gasket or weld seal. The moral: Always specify chrom-moly alloy or stainless steel as body material where saturated water is encountered or where you suspect steam to condense (practically any process steam line that gets shut down at night or on weekends). For the third problem, specify hardened trim or use angle valves. There are, in my book, only two cases when you should use an angle valve: one, for flashing condensate at the inlet of a condenser and two, for erosive fluids. In both cases, the flow should be in the direction to close the plug (check actuator for stability). For erosive service, use a reduced trim and a long, straight, (ideally) oversized downstream pipe to avoid particle impingement. The fourth problem can sometimes be cured by reversing the flow direction; otherwise, increase the stem diameter or reduce the weight (mass) of the valve plug. The fifth problem can be dealt with when selecting materials with the aid of a good handbook (see Chapter 11). For hints on solving the problem of excessive stem packing leakage, see Chapter 12.

Rotary control valves fall into two broad categories:

- Those designed exclusively for modulating control
- On-off valves modified by the addition of an actuator and positioner (also called "automated valves")

In the first category are many features commonly seen in globe valves such as adjustable stem packing, the option of a hard-faced trim, extension bonnets, and, most important, a backlash-free connection between the rotating "closure member" and the actuator stem. Some more refined types even have quick-change trim and anticavitation along with low-noise trim inserts (see Figure 3-1).

The second category, while admittedly less expensive, can be perilous. Here, the main concern should be the backlash or play between:

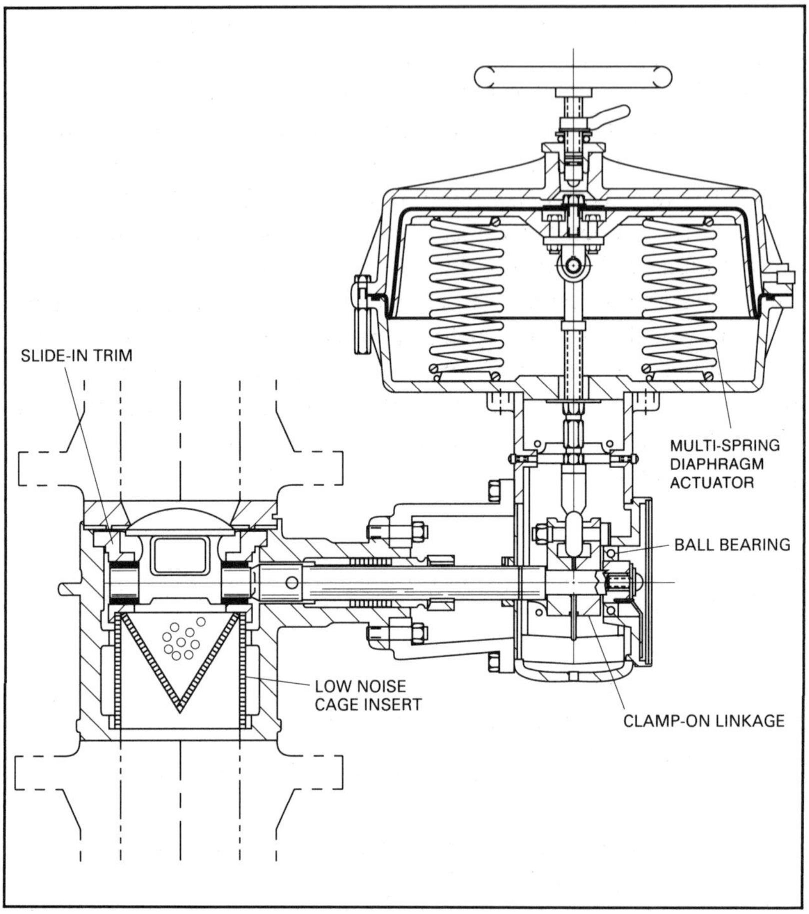

Figure 3-1. Example of a rotary valve designed especially for control purposes. Note field-replaceable quick-change trim and optional low-noise insert. (Courtesy of CASHCO, Inc.)

- the ball (in case of a ball valve) and the valve stem tang,*
- the valve stem and the actuator stem tang, and
- the positioner stem and the actuator stem tang.

*The flattened portion at the end of a valve stem.

Assuming each has a 1 in. stem diameter and 0.015 in. of tang clearance, a little mathematics will prove that the actuator will have to move 2.5° or 2.8% of 90° travel before the ball will move in the opposite direction. This is acceptable in some applications, but for critical control, it would not be. Another area of concern is "break-away" friction. This is overcoming the static friction of the ball or disk in order to open the valve, a problem especially irksome on so-called "high-performance" butterfly valves. Upon start-up, it may happen that the actuator is so charged up with air pressure that the valve may fly almost wide open (after breaking out of the seat) before the positioner can catch its breath. A good rule to remember is to limit the valve travel of the ball and high-performance butterfly valve to 15% of travel; i.e., make sure your flow rate does not require a C_v that is less than that at 15% opening.

CRITERIA FOR VALVE SELECTION: A SUMMARY

Ability To Control Well (Performance)

Does the valve have low dead band, sometimes confused with hysteresis?

Will it be able to operate without a positioner (also a cost factor)?

Does it have adequate rangeability?

Is the flow characteristic acceptable?

How is its dynamic performance (frequency response, time constant)?

Installation and Effects on the Environment

Is the face-to-face such that it can be replaced with another make?

How heavy is it? Are pipe supports needed? Is it too tall?

Does it shut off tightly (if required)?

Is it going to be noisy under the given flow conditions?

Would packing leakage produce a hazard?

Is there adequate air pressure (in case of cylinder actuators)?
Can it withstand the corrosive effects of fluid *and* environment?

Maintainability and Long-Term Cost Control

Is it easy to repair?
What are the costs of spare parts?
Can the positioner handle dirty air (if applicable)?
How is the actuator protected from corrosion?
Is the stem packing field-replaceable?
Does it need to be drained?

Only after reviewing these criteria and deciding on the proper valve type should the subject of cost be considered.

4

THE SELF-ACTING REGULATOR, WHY NOT?

The self-acting regulator is a relatively inexpensive, complete control system that combines the process sensor, the controller, and the control valve all in one. Sadly, the market share of this device is dying out from sheer neglect. Properly utilized, regulators could replace control valves in at least 25% of all control loops. However, they are seldom specified in a process loop due to the instrument engineer's unfamiliarity with them. Yet almost every valve positioner comes with its own air set, which is nothing but a self-contained air pressure regulator.

The advantages of regulators,[1] besides cost, are: an excellent frequency response (in virtually milliseconds), very good rangeability, space saving, usually have no stem leakage, and they do not require air or electricity to operate. Here, the process fluid itself, be it the differential pressure across the valve or the temperature difference, provides the necessary energy to operate the valve.

The disadvantages are: fixed proportional band (typically 25% for direct and 5% for pilot-operated models), no reset action, limited sizes and pressure ratings, and limited choice of materials and end connections. The maximum C_v rating is usually about one half that of a control valve, which is still adequate for line-size installations.

An important safety consideration is the fact that spring-loaded, pressure-reducing types of regulators will open fully if the operating diaphragm fails! It is, therefore, important to make sure that the downstream piping can withstand upstream pressure conditions; otherwise, a downstream relief valve that can handle the maximum flow of the regulator must be installed. Spring-loaded, backpressure regulators will close upon diaphragm failure, and a similar backup for protection of the upstream piping system is advised. Nevertheless, when the application calls for simple control of a blanketing gas in a tank, the steam tracing of a crude oil pipe, or other secondary loop applications, regulators should be given serious consideration.[2] They may also be the answer to applications where very fast reaction to a process upset is a must.

References

1. Moore, R. L., "The Use and Misuse of Pressure Regulators," *Instrumentation Technology*, pp 52–56, March (1969).
2. Baumann, H. D., "Regulator or Control Valve," *ISA Journal*, pp 51–52, December (1965).

5

VALVE SIZING MADE EASY

VALVE SIZING

This activity can be defined as picking the right size valve for the job or, in certain instances where a valve size is predestined, the correct trim size.*

To help matters, about 30 years ago some clever soul invented a valve sizing coefficient called C_v that combined the flow area of the valve orifice, the contraction coefficient, and the head loss coefficient all in one; thus,

*Do not read on if you have a computer fully programmed with ISA-S75.01 equations.[1] But then again, you may want to know what you are doing.

$$C_v = q/\sqrt{\Delta P/G_f} \tag{5-1}$$

$$q = C_v\sqrt{\Delta P/G_f} \tag{5-2}$$

where q is the flow rate in U.S. gallons per minute, G_f is the specific gravity of a liquid at a given temperature (water at 60°F has a G_f of one), and ΔP is inlet minus outlet pressure in psi.

Hence, one C_v equals the flow of one U.S. gallon per minute of water at 60°F and under a pressure drop across the valve of 1 psi. Now the same valve will flow 10 gpm at a pressure drop of 100 psi since the flow is proportional to the square root of ΔP, thus following the laws of fluid mechanics. This equation is worthwhile memorizing since you will use it on at least 80% of all sizing problems for liquids. Unfortunately, not all in life is so simple; therefore, we have to recognize two limitations in the use of the above equations.

First, they only apply to turbulent flow, i.e., Reynolds numbers above about 5000. However, you can ignore this limit on required C_v numbers above 0.1 and when the kinematic viscosity is below 40 centistokes (90% of all liquids are). There will be more on this subject—required reading only if you control tar or equally sticky substances and polymers.

Second, in some conditions funny things occur inside the valve that seem to block the flow passages. What really happens is that, at a certain ratio of pressure drop to inlet pressure defined by a pressure recovery factor F_L,[1] the flow becomes choked at the orifice (I originally called F_L the "critical flow factor,"[2] C_f, still used as such by many). In the case of liquids, this is caused by the evaporation of some of the liquid (also see "Cavitation" in Chapter 12), and, in the case of gases, the onset of sonic velocity. We could not care less about all this were it not for the fact that now the flow is independent of the downstream pressure P_2.

To avoid embarrassing errors, you should first start to determine if you have critical (choked) flow or not (this also helps to discover any cavitation or noise problems you may have to worry about).

For Liquid Service

SUBCRITICAL FLOW	CRITICAL FLOW	
When ΔP is less than $F_L^2(\Delta P_s)$	When ΔP is more than $F_L^2(\Delta P_s)$	ΔP_s = Maximum ΔP for sizing Use: $P_1 - P_v$ when outlet pressure is higher than vapor pressure. Use: $P_1 - \left(0.96 - 0.28\sqrt{\frac{P_v}{P_c}}\right) P_v$ when outlet pressure is equal to or lower than vapor pressure.

Volumetric Flow

$$C_v = q\sqrt{\frac{G_f}{\Delta P}} \qquad C_v = \frac{q}{F_L}\sqrt{\frac{G_f}{\Delta P_s}}$$

Flow by Weight

$$C_v = \frac{W}{500\sqrt{G_f \Delta P}} \qquad C_v = \frac{W}{500F_L\sqrt{G_f \Delta P_s}}$$

G_f = specific gravity @ flowing temperature (water = 1 @ 60°F)

W = flow in pounds per hour

$\Delta P = P_1 - P_2$

q = liquid flow rate, U.S. gpm

P_c = pressure at thermodynamic critical point, psia—water is 3206 psia

P_v = vapor pressure of liquid at flowing temperature, psia

For Gas and Steam Service

Surprising as it may sound, the basic C_v* equation can also be applied to gases and steam. The difference here is that the density of

*There is evidence that the C_v number on certain valve types such as angle valves, certain rotary valves, and double-ported globe valves is different between liquid and gases, but we shall ignore this.

the gas changes with ΔP, and, since this is a gradual process, the relationship $\sqrt{\Delta P}$ to flow is no longer linear but curved. To its credit, the ISA Sizing Committee for S75.01 has labored hard to predict the exact shape of this curve with the aid of an expansion factor,[3] Y, where

$$Y = 1 - (x/3F_k X_T) \tag{5-3}$$

Note that Y cannot exceed 1 and that it cannot be less than 0.67 since the latter is exactly the density of the gas, i.e., 67% of the inlet density when the valve is choked (when it sees sonic velocity in the valve orifice).

In the above equation,

x = $\Delta P/P_1$ (Note: P_1 or P_2 is *always* absolute pressure, i.e., gage psi plus 14.7)

F_k = the ratio of the specific heat k of a gas (air = 1.4), F_k = k of gas/1.4

X_T = pressure drop ratio. This is the $\Delta P/P_1$ at which there is no longer any flow increase. Note that X_T includes some flow increase *after* the onset of choked flow due to changes in the size of the flowing jet (vena contracta).

Now the required flow coefficient becomes:

$$C_v = q\sqrt{G_g T_1 Z/x}/1360 P_1 Y \tag{5-4}$$

or

$$q = 1360 C_v P_1 Y/\sqrt{G_g T_1 Z/x} \tag{5-5}$$

where:

q = flow in scfh

G_g = gas specific gravity (air = 1), also ratio of molecular weight of gas to that of air (air = 28.9)

T_1 = absolute upstream temperature, °R(°F + 460)

Z = compressibility factor (Ignore if your pressure never exceeds 2000 psia; otherwise, consult handbook tables.)

Now, if this begins to tax the capabilities of your hand-held calculator, you may ignore all of the above and use the simplified version below.[4] (The penalty is a maximum sizing error of ≈8% at the transition from non-choked to choked flow and not worth the trouble.)

SUBCRITICAL FLOW	CRITICAL FLOW
When ΔP is less than $F_L^2(P_1/2)$	When ΔP is more than $F_L^2(P_1/2)$

Volumetric Flow

$$C_v = \frac{q}{963}\sqrt{\frac{G_g T}{\Delta P(P_1 + P_2)}} \qquad C_v = \frac{q\sqrt{G_g T}}{834 F_L P_1}$$

Flow by Weight

$$C_v = \frac{W}{3.22\sqrt{\Delta P(P_1 + P_2)G_g}} \qquad C_v = \frac{W}{2.8 F_L P_1 \sqrt{G_g}}$$

For Saturated Steam

$$C_v = \frac{W}{2.1\sqrt{\Delta P(P_1 + P_2)}} \qquad C_v = \frac{W}{1.83 F_L P_1}$$

For Superheated Steam

$$C_v = \frac{W(1 + 0.0007T_{sh})}{2.1\sqrt{\Delta P(P_1 + P_2)}} \qquad C_v = \frac{W(1 + 0.0007T_{sh})}{1.83 F_L P_1}$$

where:

C_v = valve coefficient
F_L = pressure recovery factor
G_g = gas spccific gravity (air − 1.0)
P_1 = upstream pressure, psia
P_2 = downstream pressure, psia
ΔP = pressure drop $P_1 - P_2$, psi
q = gas flow rate at 14.7 psia and 60°F, scfh
T = flow temperature, °R (460 + °F)
T_{sh} = steam superheat, °F
W = flow rate, pounds per hr

NOTE: When a valve is installed between reducers, use $C_v F_p$ instead of C_v in capacity tables (see section on "Pipe Reducers").

OTHER METHODS

$$C_v = \frac{C_g}{40F_L}; \qquad C_v = \frac{C_s}{2F_L}; \qquad X_T = 0.84F_L^2$$

$$F_L = \sqrt{K_m}; \qquad F_L = C_1/40$$

NOTE: X_T is not strictly related to F_L. While X_T gives you the maximum gas flow, which may happen at orifice pressure ratios P_1/P_0 of 3 : 1 or more, i.e., above sonic velocity, F_L^2 tells you the onset of sonic velocity, i.e., where $P_1/P_{\text{orifice}} \approx 2$, and $P_1/P_{2\,\text{sonic}} = P_1/(P_1 - 0.5F_L^2)$.

WHAT ARE THE RIGHT FLOW CONDITIONS?

Sizing the C_v number to the fifth place after the decimal point is all well and good, but only if you have the right process conditions. First of all, the valve should be sized to control the maximum flow rate your process is designed for. But what is this exactly? How about emergency conditions? On the other hand, should the valve fail open, would the resultant flow be more than the downstream pressure relief valve could handle? Then again, what is the correct inlet pressure? Can you use the head pressure of the pump from the manufacturer's published curve? (If you do, use the head pressure corresponding to the maximum flow the valve has to pass, and don't ignore the static head at the pump's location.) What is the head loss or pressure drop across a heat exchanger next to the valve? How about line losses? All these are questions that beg only vague responses from your process design people. They don't know either and have to rely on the input from other vendors. The result is usually guesswork with ample safety factors thrown in. Therefore, don't be too concerned about the accuracy of sizing equations. Ninety-nine out of one hundred times, the valve is too large for the job anyhow (see Chapter 7 on equal percentage trim). The remaining one percent that is undersized comes about because someone made a decimal point error in the C_v calculation.

If you want to be clever and save your company energy and money in the bargain, have the process people *ask you* for the valve pressure drop when they design the system (see Appendix A). Now you are in the driver's seat. Here are the very scientifically derived rules of thumb you should memorize: For ΔP valve sizing, use the higher of 5% of the total system pressure (i.e., static head plus pump head at maximum flow, boiler pressure, etc.), or 5 psi for rotary control valves, or 10 psi for globe valves.

You may wonder why I picked 5 psi for rotary control valves. Well, it just so happens that the flow rate (either liquid or gas) generated by a 5 psi pressure drop through a wide open control valve with a C_v divided by the valve diameter squared of 17* is all that a pipe (the same size as the valve) can handle, allowing for the customary maximum line velocities of 15 ft/s for liquids or 150 ft/s for gases[5] as you can see from Figure 5-1. Higher ΔP's mandate a valve smaller than the pipe size and where reducers are required, or when reduced trim is selected.** In any case, you are throwing away dollars for pump, compressor, or boiler horsepowers if you assign too much pressure drop to the valve.

Of course, the above trick can be used in reverse. You may be requested to replace an old steam reducing valve, vintage 1948. All the records are lost, and the only thing you know is the boiler pressure and the valve and pipe size. Say the boiler pressure is 85 psig (100 psia) and the pipe size is $1\frac{1}{2}$ inches. First, look up the maximum flow rate of a $1\frac{1}{2}$ in. steam pipe at, say, 50 psig downstream pressure. You will find it to be approximately 1000 lbs/hr (see Table A-2 in Appendix A). Next, you assume the flow to be choked, i.e., P_2 is less than one-half P_1. In any case, people didn't care much about energy savings in 1948! Now calculate the C_v required, which is about 12. So go ahead and select a $1\frac{1}{2}$ in. globe valve with a rated C_v of about 22. The worst that can happen is that the outlet pressure is 80 psig

---

*A butterfly valve at 60° opening, for example.

**For example, if a 4 in. butterfly valve is installed in a 6 in. pipe, then at the same $17 \times d^2 = 272\,C_v$, the valve pressure drop increases from 5 psi to 25 psi with 15 ft/s fluid velocity in the 6 in. pipe.

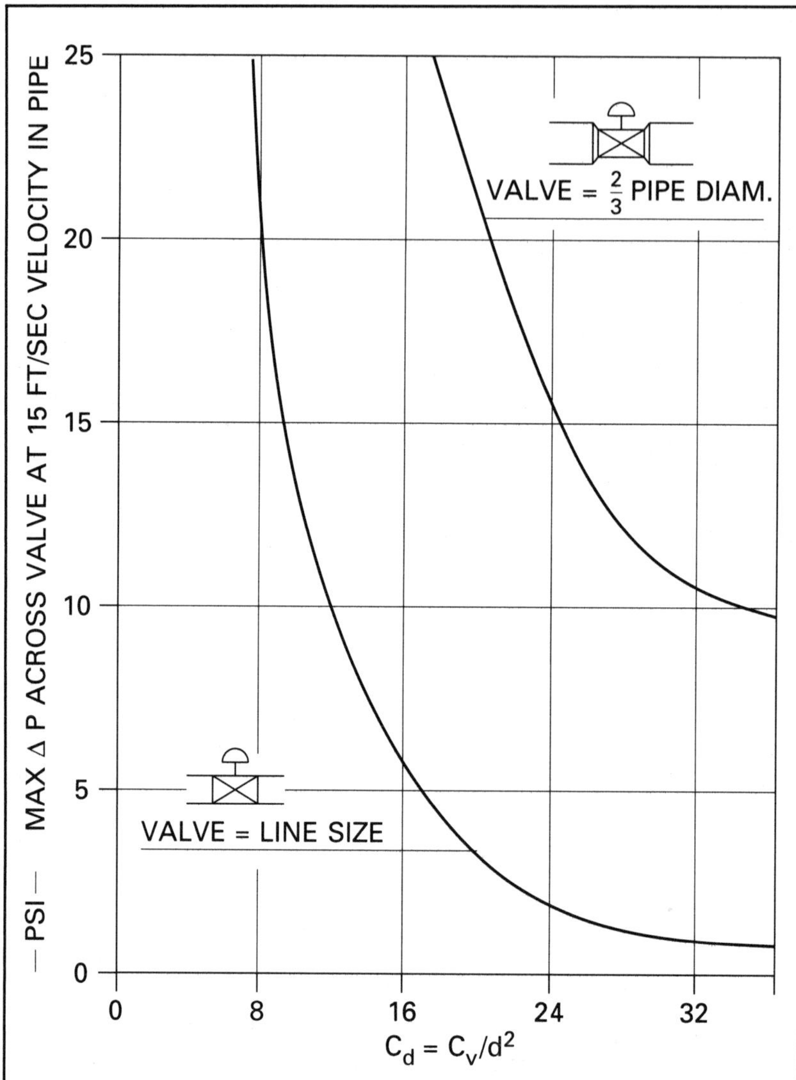

Figure 5-1. Relationship between pressure drop and specific valve flow capacity for water at 15 feet/second velocity in pipe, *d* = valve diameter (inches).

instead of below 50 psig. No problem. Now the C_v required is 16, which is still OK. Therefore, when in doubt, use a full-trim size if flow conditions are unknown except in special applications such as pH control.

PIPE REDUCERS

One of the competitive strategies of some valve manufacturers is to offer a control valve that has a much higher C_v rating than somebody else's valve. These are usually oversized globe valves with reduced flange connections, for example, a $1\frac{1}{2}$ in. flanged globe valve with a 2 in. orifice size. While there is nothing wrong with this concept (after all, you are getting more weight for your dollars), it nevertheless forces you into using pipe reducers, since with a large C_v your valve is smaller than the pipe. So one thing you have to do is evaluate the extra cost and space requirement.

You may also need reducers when you buy rotary control valves, which, due to having less torturous bodies, have inherently more C_v—although with higher ΔP's this cancels out due to low F_L numbers (choked flow is dependent on the product of C_v times F_L!).

A word of caution: Never choose a valve that is less than half the pipe size. Too much stress may be created by the movement of the pipes. The only two times I was genuinely scared in my life was (1) witnessing a live test with steam on a newly designed steam dump valve for nuclear power and (2) seeing a 1 in. steam makeup valve welded into a 10 in. schedule 80 pipe, reducing 1800 psi steam to 10 psig!

There is a good reason why I put this discussion near the end of the chapter on valve sizing. Namely, reducers do not affect you that much. Simply remember that the combined flow coefficient of the *valve and reducer* is C_vF_p, where F_p is the piping geometry factor (which I used to call R,[6] a term still utilized by one manufacturer). Some manufacturers have published tables showing the reduced $C_v(C_vF_p)$ due to the pressure losses caused by the pipe reducers. *Make sure the C_v or C_vF_p value of the valve you select is at least 10% higher than the C_v number calculated with the above equations.* For globe valves at rated C_v's of 10 d^2 or less (d is valve size in inches), F_p is practically one, so it can be ignored. Things are more critical with butterfly or ball valves where F_p can be as low as 0.5!

Here is a quick guide to F_p factors. First, divide the valve catalog C_v by the valve diameter d (inches) squared, then read F_p:

Table 5-1. F_p Values for Valves, $D/d = 2$

C_v/d^2	10	15	20	25	30	35	40	45
F_p	0.96	0.91	0.85	0.79	0.74	0.68	0.63	0.59

D = pipe diameter (inches)

CORRECTING FOR VISCOSITY

Since this is a procedure not often encountered, I put it last. Ignore this unless (a) you have a liquid with a viscosity exceeding 40 centistokes (or ≈ 40 centipoises for liquids), or (b) you need really small valves with a C_v of less than 0.1.

Similar to the reducer correction, we have to increase the C_v from that calculated with the standard sizing equations in order to make up for the additional friction caused by the stickiness of the stuff passing through the valve. This is expressed by the "valve Reynolds number factor," F_R. The valve C_v you *really* need = C_v originally calculated$/F_R$. More precisely,

$$C_{v_{\text{corrected}}} = \frac{C_{v_{\text{turbulent}}}}{F_R} \tag{5-6}$$

where:

$C_{v_{\text{turbulent}}} = q\sqrt{G_f/\Delta P}$ (see valve sizing formulas)

Re_v, (valve Reynolds number) $= 17300\, F_d q/\nu\sqrt{F_L C_{vR}}$

q = flow rate, gpm

C_{vR} = rated C_v of valve selected (may have to be increased if $C_{v\text{corrected}}$ is larger.)

F_L = pressure recovery factor
ν = kinematic viscosity, centistokes (10^{-6} m/s^2) = centipoises/G_f
F_d = valve style modifier

The F_R number varies with the internal fluid resistance (K factor) of each valve type, i.e., C_v/d^2. For graphical values of F_R, see Figure A-1 in Appendix A.

For typical values for ν, see Tables 5-2 and 5-3.

Table 5-2. Viscosities and Specific Gravities for Liquids

Liquid	Formula	G_f	T(°F)	ν(cSt)
Acetic acid	$HC_2H_3O_2$	1.05	68	1.17
Acetone	C_3H_6O	0.79	68	0.42
Alcohol, ethyl	C_2H_6O	0.79	68	1.52
Alcohol, methyl	CH_4O	0.79	68	0.74
Ammonia	NH_3	0.62	68	0.24
Aniline	C_6H_2N	1.02	60	4.31
Auto oil, SAE 30		0.94	100/210	114/11.2
Benzene	C_6H_6	0.88	68	0.74
Brine, CaCl 25%	CaCl	1.23	68	2.03
Brine, NaCl 25%	NaCl	1.19	68	1.60
Bromine	Br_2	2.93	60	0.34
Carbon dioxide	CO_2	0.84	−109	0.83
Carbon disulfide	CS_2	1.30	32	0.33
Carbon monoxide	CO	0.84	−314	0.21
Carbon tetrachloride	CCl_4	1.59	77	0.57
Castor oil		0.96	68	1027.00
Chlorine	Cl_2	1.50	−100	0.50
Citric acid	$C_6H_8O_7$	1.11	68	2.00
Dow Therm A		0.79	68	5.60
Dow Therm A		0.79	200	1.07
Dow Therm E		0.79	68	1.93
Dow Therm E		0.79	200	0.53
Ethylene glycol	$(CH_2OH)_2$	1.12	68	16.59
Fluorine	F_2	1.51	−306.4	0.16
Formaldehyde	H_2CO	0.82	60	0.35

Table 5.2. *(Continued)*

Liquid	Formula	G_f	T(°F)	ν(cSt)
Freon 12	CCl_2F_2	1.31	77	0.47
Fuel oil, #6 (Bunker C)		0.98	85	4080.00
Fuel oil, #6 (Bunker C)		0.97	110	1100.00
Gasoline		0.71	60	0.86
Glycerine (100% glycerol)	$C_3H_8O_3$	1.26	68	920.00
Glycol (ethylene glycol)	$C_2H_6O_2$	1.11	60	17.90
Hydrochloric acid (31.5%)		1.64	60	1.52
Hydrogen	Hz	0.074	−422	0.011
Hydrogen chloride	HCL	0.91	68	2.67
Hydrogen fluoride	HF	0.97	68	0.13
Hydrogen sulfide	H_2S	0.79	60	
Isopropyl alcohol	C_3H_8O	0.79	68	2.94
Linseed oil		0.93	60	55.90
Kerosene		0.82	85	2.99
Mercury	Hg	13.57	60	0.12
Naphthalene	$C_{10}H_8$	1.15	68	398.00
Nitric acid (40%)	HNO_3	1.25	68	1.25
Nitrogen	N_2	0.85	−320	0.19
Oxygen	O_2	1.20	−297	0.17
Phosphoric acid	H_3PO_4	1.15	68	1.85
Propylene glycol (60%)	$C_3H_8O_2$	1.04	68	8.99
Sodium hydroxide	NaOH	1.27	68	6.20
Starch (4% Soln)	$(C_6H_{10}O_5)x$	1.02	100	100.00
Sulfuric acid (100%)	H_2SO_4	1.83	68	13.91
Toluene	C_7H_8	0.87	68	0.68
Trichloroethylene	C_2HCl_3	1.46	80	0.39
Turpentine		0.87	60	1.71
Water, fresh	H_2O	1.000	60	1.00
Water, sea (4% by wgt)		1.03	60	1.05

Table 5-3. Viscosities and Specific Gravities for Gases at Atmospheric Pressures

Gas	Formula	G_g	T(°F)	ν(cSt)
Acetic acid	$HC_2H_3O_2$	2.07	246	5.75
Acetone	C_3H_6O	2.01	212	4.91
Acetylene	C_2H_2	0.907	32	7.99
Air	N_2O_2	1.000	64	15.10
Alcohol				
Butanol, butyl	$C_4H_{10}O$	2.56	242	10.94
Ethanol, ethyl	C_2H_6O	1.59	212	7.20
Methanol, methyl	CH_4O	1.11	152	11.77
Propanol, propyl	C_3H_8	2.07	212	4.76
Ammonia	NH_3	0.596	68	13.71
Argon	A	1.379	68	13.38
Benzene	C_6H_6	2.70	58	2.23
Bromine	Br_2	5.52	55	2.21
Carbon dioxide	CO_2	1.5290	68	8.06
Carbon disulfide	CS_2	2.63	68	2.88
Carbon monoxide	CO	0.97	59	14.51
Carbon tetrachloride	CCl_4	5.31	170	3.65
Chlorine	Cl_2	2.486	68	4.44
Cyanogen	C_2N_2	1.806	63	4.51
Ethane	C_2H_6	1.049	63	7.08
Ether	$(C_2H_5)_2O$	2.56	58	2.28
Ethyl chloride	C_2H_5CL	2.220	32	3.27
Ethylene	C_2H_4	0.975	68	8.60
Fluorine	F_2	1.31	68	13.98
Freon 11	CCl_3F	5.04	68	1.77
Freon 12	CCl_2F_2	4.2	68	2.46
Helium	H_e	0.138	−431	0.89
Helium	H_e	0.138	68	117.06
Hydrogen	H_2	0.0695	−431	0.37
Hydrogen	H_2	0.0695	68	104.90
Hydrogen bromide	HB_r	2.819	66	5.35
Hydrogen chloride	HCl	1.268	64	9.32
Hydrogen cyanide	HCN	0.93	68	8.95
Iodine	I_2	8.76	338	2.93
Isopropyl alcohol	C_3H_8O	2.08	212	5.55
Krypton	KR	2.89	59	6.96
Mercury	Hg	6.93	523	11.05

Table 5.3. *(Continued)*

Gas	Formula	G_g	T(°F)	ν(cSt)
Methane	CH_4	0.554	68	15.41
Methyl acetate	$C_3H_6O_2$	2.56	212	4.14
Methyl chloride	CH_3Cl	1.74	59	4.89
Methyl iodide	CH_3I	4.47	811	14.66
Neon	NE	0.69	68	37.53
Nitric oxide	NO	1.04	68	15.01
Nitrogen	N_2	0.97	81	15.66
Nitrosyl chloride	ClNO	2.26	59	4.12
Nitrous oxide	N_2O	1.52	80	8.33
Oxygen	O_2	1.10	66	15.23
Pentane (n)	C_5H_{12}	2.49	77	2.30
Propane	C_3H_8	1.52	64	4.32
Propylene (propene)	$C_2H_4(CH_2)$	1.45	62	4.71
Sulfur dioxide	SO_2	2.21	69	4.73
Saturated steam (30 psia)	H_2O	0.62	250	12.20
Toluene	C_7H_8	3.18	68	1.83
Xenon	XE	4.53	68	4.15

For pressures other than atmospheric, multiply listed ν by $14.7/P_1$, where P_1 = actual inlet pressure in psia.

Typical values for F_d (also refer to Table 12-2):

Globe valves ($C_v > 0.1$)
Single-seated, parabolic plug = 0.46
Double-seated, parabolic plug = 0.32
Butterfly valve = 0.57
Segmented ball valve = 0.80
Eccentric rotary plug valve = 0.42

NOTE: The above F_d values do not currently agree with ISA-S75.01-1985.

Example: Bunker C Oil @ 120°F, P_1 = 100 psig, P_2 = 75 psig, $G_f = 0.97$, q = 26 gpm, ν = 750 centistokes,
$C_{v_{\text{turbulent}}} = 26\sqrt{0.97/25} = 5.12$

Select: $1\frac{1}{2}$ in. globe valve, parabolic plug, rated $C_v = 28$,* $F_L = 0.9$, $F_d = 0.46$ at rated C_v (see Table 12-2); $C_v/d^2 = 12.4$; F_R from Figure A-1 is 0.42 for $\text{Re}_v = 55$.

$$\text{Re}_v = \frac{17300 \times 26 \times 0.46}{750\sqrt{0.9 \times 28}}$$

$$C_{v_{\text{corrected}}} = \frac{5.1}{0.42} = 12.1 \qquad (\text{OK, } 43\% \text{ rated } C_v)$$

Viscosity Conversion: If viscosity is given in centipoises, divide centipoise (μ) by specific gravity (G_f) of liquid to obtain centistokes. If viscosity is given in Saybolt Seconds Universal (t), approximate centistokes by multiplying Saybolt seconds by 0.2 (example: $t = 89$, $\nu = 18$).

Small Flow Valve Sizing: Now for small flow valves ($C_v \leq 0.1$), we calculate the valve Reynolds number Re_v as before, except we use different F_d factors,[7] namely:

- 0.7—for splined plugs
- 0.3—for short travel diaphragm valves with a flat seating surface
- $F_d = 0.1 \times \sqrt{C_v}/(\text{plug diameter})$, for close clearance tapered plugs

Example: For a $\frac{1}{8}$ in. diameter close clearance plug with $C_v = 0.05$:

$$F_d = 0.1\sqrt{0.05}/0.125$$

$$= 0.18$$

*Valve size may be increased if calculations show valve is too small.

For liquids:

$$\mathrm{Re}_v = \frac{17300 \times F_d \times q}{\nu\sqrt{C_v F_L}} \tag{5-7}$$

For gases:

$$\mathrm{Re}_v = \frac{2153 \times F_d \times q}{\nu\sqrt{C_v F_L}} \tag{5-8}$$

where q = flow in scfh @ 14.7 psia and 60°F.

Now calculate F_R: If $\mathrm{Re}_v \geq 130$,

$$F_R = \sqrt[n]{\mathrm{Re}_v/10000} \qquad \text{(Note: } F_R \text{ max} = 1)$$

where $n = 1 + \log(\mathrm{Re}_v)$.

If $\mathrm{Re}_v \leq 130$, then

$$F_R = \sqrt{\mathrm{Re}_v/2100}$$

For example:

Fluid: CO_2 gas at 85°F (T = 545°R)

P_1 = 36.3 psia

P_2 = 24.7 psia

ν = kinematic viscosity = 8.65 cSt @ 85°F and 14.7 psia (NOTE: Use ν at atmospheric pressure if you use q in scfh.)

G_g = specific gravity = 1.516

C_v = valve sizing coefficient (selected) = 0.025

F_L = liquid pressure recovery factor = 1 (from mfr.'s catalog)

F_d = valve style modifier = 0.16 (from mfr.'s catalog)

q = desired max. flow rate = 15 scfh at 85°F and 14.7 psi absolute

T = 460 + 85 = 545°R

From Equation 5-8,

$$\mathrm{Re}_v = \frac{2153 \times 0.16 \times 15}{8.65 \times (1 \times 0.025)^{0.5}}$$
$$= 3778$$

$$F_R = 4.6\sqrt{\frac{3778}{10000}};\ n = 1 + \log(3778) = 4.57$$
$$= 0.808$$

Check if the selected valve C_v of 0.025 is sufficient to handle the flow of 15 scfh.

From the turbulent gas equation for subcritical flow* and modified by F_R,

$$C_v = q\sqrt{G_g T/\Delta P(P_1 + P_2)}/963\ F_R$$
$$= 15\sqrt{1.516 \times 545/11.6 \times 61}/963 \times 0.81$$
$$= 0.02;$$

so the selected valve C_v of 0.025 is OK.

METRIC UNITS

Normally, it does not bother you that the rest of the world thinks in metric terms. However, an occasion may arise when you will have to do a project for an overseas plant, and you will be confronted with liquid flow rates of cubic meters per hour instead of gpm. Luckily, the C_v numbers work equally as well in Europe.

---

*Remember, there is no sonic velocity if flow is nonturbulent; therefore, critical flow equations don't apply.

For example:

$$C_v = 1.17\ q/\sqrt{\Delta P/G_f} \tag{5-9}$$

where:

q = m^3/hr of liquid
ΔP = bar
G_f = specific gravity (water @ 60°F = 1).

As a matter of fact, in Germany they go so far as to eliminate the 1.17 constant and then call the flow coefficient K_v; hence, $K_v = 1.17\ C_v$, or $C_v = 0.85\ K_v$.

Complications arise only if you are confronted by newtons per square meter, also called pascal or Pa. To get back to the more familiar bar, divide Pa by 100,000 or one bar = 100,000 Pa, and one bar = 14.5 psi. One kilopascal (kPa) = 0.15 psi. There is also the less often used kg/cm^2, which is 0.98 bar or 14.2 psi.

Remember also, that degrees Celsius is used abroad where °F = (°C × 9/5) + 32, and in absolute terms, °K = 273 + °C.

Converting flowing units:
1 m^3/hr liquid = 4.4 gpm
1 m^3/hr gas = 35.3 ft^3/hr
kg/hr = 2.2 lbs/hr

You are now fully equipped to work abroad.

WHAT SIZE VALVE TO CHOOSE

After you have labored hard to select the final required valve C_v suitable for maximum flow requirements, and having considered the F effects (F_L, F_p, F_R), you are now confronted with the final choice—the valve size.

If the pressure drop is low, say, 5 to 15 psi, it is always safe to start with a line size valve. Keep in mind that the person in the Piping

Design Department has already sized the downstream or upstream pipe to pass the design flow rate at a reasonable velocity. Incidentally, valve outlet velocity should not bother you unless there could be a noise problem with high pressure gas or steam reduction (see the aerodynamic noise discussion in Chapter 12), or where you have flashing liquids (if the downstream pipe is too long, your valve will cavitate due to downstream pressure buildup caused by high mixed-phase velocity). This then leaves only the C_v rating of the valve. In case you have not yet selected a certain brand because you are partial to the color of the valve, i.e., green or red, here is a hint:

Table 5-4. Typical Flow Coefficients at Rated Travel

Type	C_v	F_L	C_vF_L	C_vF_p $D/d = 1.5$		C_vF_p $D/d = 2$	
Single-seated globe valve parabolic plug, flow-to-open	10	0.9	9	9.8	(4.3)	9.6	(2.4)
Double-seated globe valve	12.5	0.9	11.3	12	(5.3)	11.7	(2.9)
High capacity cage valve to size 3 in. only	14	0.9	12.5	13.5	(6.0)	12.9	(3.2)
Eccentric rotary plug valve	12	0.85	10	11.6	(5.1)	11.3	(2.8)
Butterfly valve 60° open	17	0.68	11.5	15.8	(7.0)	15	(3.8)
Butterfly valve 70° open	27	0.57	15.5	23	(10.2)	20.8	(5.2)
Fluted butterfly valve @ 70°	25	0.7	17.5	21.8	(9.7)	19.8	(5.0)
Contoured ball valve @ 90°	25	0.6	15	21.8	(9.7)	19.8	(5.0)
Angle, venturi type, flow-to-close	22	0.5	11	19.8	(8.8)	18.3	(4.6)

The C_v values are per *valve* diameter (inch) squared. The C_v values in parentheses are per *pipe* diameter (inch) squared. For example: A 6 in. eccentric rotary plug valve has a wide open C_v flow coefficient of $6^2 \times 12 = 432$. However, when installed between reducers in a 12 in. pipe, the available flow coefficient is only $6^2 \times 11.3 = 407$ (see column C_vF_p for $D/d = 2$), or $12^2 \times 2.8 = 403$ using the figures in parentheses.

The above values are conservative and vary somewhat from size to size. You may notice that when high capacity valves are used under choked flow conditions (such as steam pressure reduction), they all tend to lose their high C_v advantage, as you can see from the C_vF_L column!

Starting out with a line size valve, divide your desired maximum C_v value by 0.8. First, this will give you a 10% safety factor and second, it will guard against the $\pm 10\%$ tolerance in the C_v rating of the valve given by the manufacturer. Now divide this new C_v number by the pipe size squared. For example: From the flow data, you require a maximum $C_v = 123$. Your line size is 4 inches. First divide 123 by 0.8 to obtain 154. Now divide this by (4 in.)^2. $154/4^2 = 9.6$ C_v/d^2.

From the data given in Table 5-4, you have the following choices:

- A 4 in. single-seated globe valve, $C_v/d^2 = 10$; C_v rated $= 10 \times 4^2 = 160$.
- A butterfly valve at 70° opening (if other considerations such as leakage, noise, etc., permit). Here a 3 in. butterfly valve installed in a 4 in. pipe has a flow capacity higher than 10.2 $\times D^2$ (see figures in parentheses), i.e., more than $4^2 \times 10.2 = 163$ C_v.

Having thus determined the size (and type) of valve, you now should check that you can meet the minimum required flow coefficient that your process requires. Here you have to go to the manufacturer's C_v tables (usually printed in increments of 10% travel). Look up the approximate valve travel when the valve is operating at your minimum C_v. This travel should not be less than 5% or the percentage of the valve dead band (with tight packing).* The only exceptions are certain soft seat plugs that "dip" through their seal ring, ball valves, elastomer-lined butterfly valves with angle-seated vanes, or certain diaphragm or pinch valves where there is no "hard" seating action. If a standard valve forces you to throttle below 5% travel, then your only choice is to use a larger and a smaller valve in parallel and split-range the actuator signal or, better yet, use two separate signals from your computer so that the small valve will open first, and then the larger valve starts when the small valve is about 80% open.

In summary:

- Choose a valve that has a rated C_v at least 20% larger than your maximum requirements.
- See that the minimum required C_v occurs *above* 5% of the valve travel.

ADJUSTABLE TRAVEL—ADJUSTABLE C_v

As mentioned previously, one of the vexing problems confronting the instrument engineer is the fact that the final required flow capacity that the selected valve needs is sometimes less than half of what was specified. The result is an effective signal span of less than 40% (4 – 10.4 mA) in the case of a linear plug characteristic and less than 75% (4 – 16 mA) with an equal percentage flow characteristic.

Normally, the controller or the computer does not have a problem controlling the loop with a reduced signal span, since the signal resolution is still about one in one thousand for 50% of signal span. However, the minimum controllable flow at 5% travel is higher (as a percentage of the max flow) and worse, the control valve dead band is more than twice as much as a percentage of the usable signal span, thereby affecting loop stability!

It would, therefore, be nice if we could reduce the rated valve

*You may have to use 10% if the 5% travel C_v is not known.

C_v after start-up, hopefully with the valve staying in the line. This is preferable to sending the valve back to the maintenance shed for installation of a reduced trim.

Well, valves are now available where you can adjust the rated C_v within reasonable limits, and it may be useful to become familiar with them. The illustration in Figure 5-2 shows such a valve in a

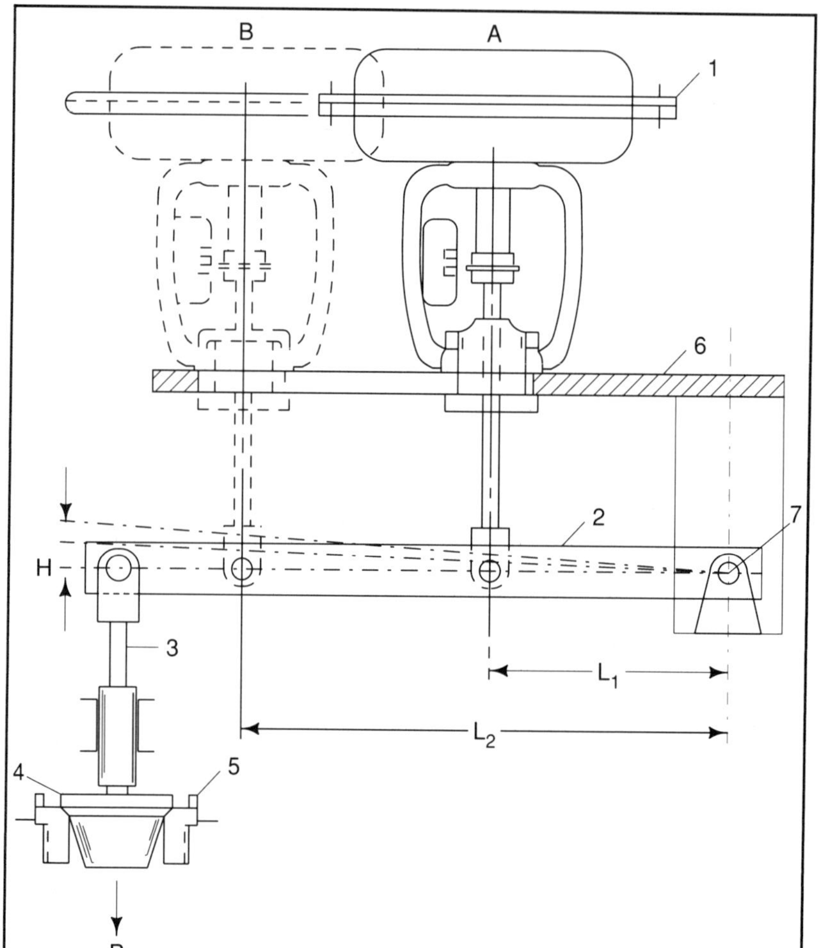

(A) **SCHEMATIC VIEW**

Figure 5-2. Control valve with equal percentage characterized plug and adjustable travel.

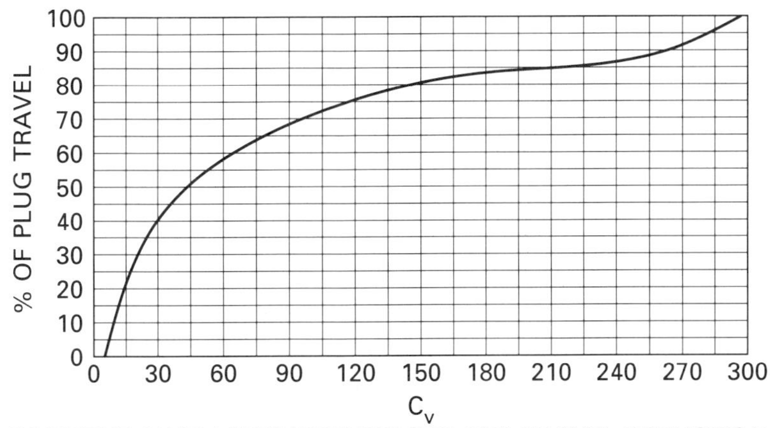

INHERENT CHARACTERISTIC BEFORE 22% TRAVEL REDUCTION
(B)

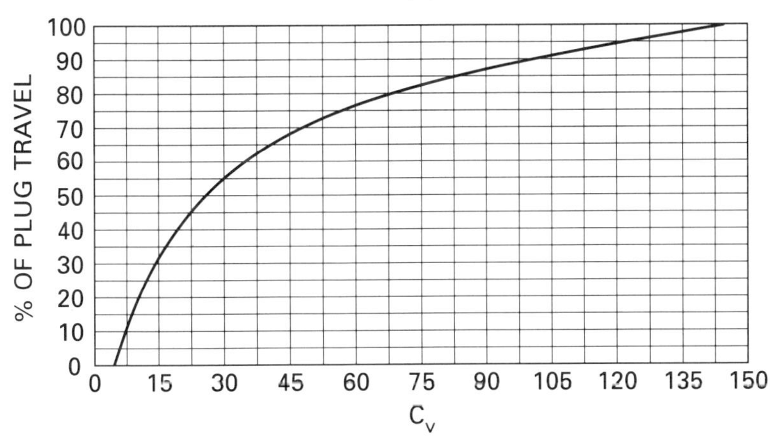

INHERENT CHARACTERISTIC AFTER 22% TRAVEL REDUCTION
(C)

Figure 5-2. *(Continued)*

schematic way to demonstrate the principle of adjustment in flow capacity. Here, a conventional spring-diaphragm actuator (1) moves a linkage (2) up and down, which, in turn, connects to a valve stem (3). The stem, in turn, moves a valve plug (4) against a seat ring (5). In the normal actuator position, A, the valve plug travels the maximum distance, H, which in turn yields the maximum valve C_v, say 300.

With an equal percentage plug contour, you get a typical "inherent flow characteristic" as shown Figure 5-2(B). Now assume that after installation we find the valve needs a flow coefficient of only 105 (equivalent to 70% of the present travel). To increase the usable signal span (= actual travel) to, say, 90% from the actual 70% (allowing a 10% safety margin), we can now relocate the actuator (1) along mounting bracket (6) to a new position, B.

Owing to a greater distance L_2 (not to scale) from the pivot point (7), the excursion of the lever (2) is reduced, resulting in a new travel H of 78% of the travel we got at actuator position A. Assuming H was 2 in. at position A, we now get 1.56 in. at B. The equal percentage characteristic at this point yields a C_v of 150, as you can see from Figure 5-2(B). With the reduced travel, we can now "spread" the curve over the available 100% signal span or over 1.56 in. plug travel. The result is shown in Figure 5-2(C).

To help you understand how this graph was developed, assume you stroke the plug to 60% travel in actuator position B. Reducing the travel now by 22% (to 78% of 2 in. or, 1.56 in. travel) will yield a new plug percentage travel of $0.78 \times 0.6 = 0.468$ or 46.8%. The actual C_v of the plug at this point is about 37 (from Figure 5-2(B)). However, since 46.8% of the actual plug travel corresponds to 60% of the actuator travel in position B, you have to plot the 37 C_v at the 60% travel position (equivalent to 60% actuator level, or 60% of signal) in order to construct the curve in Figure 5-2(C).

It is interesting to note that when comparing the two curves at 10% travel, we find a C_v value of 10 for the full travel plug but only 7.5 C_v for the reduced travel plug, yielding a somewhat wider, useful rangeability in this case ($105\ C_v/10\ C_v = 10.5:1$ over $105\ C_v/7.5\ C_v = 14:1$!).

Of greater significance is the fact that by locating the actuator (1) closer to the valve stem (3), we greatly increased the available plug force, P, in the present example, by $(1/0.78) = 1.28$ times. This not only complements the typical increase in pressure drop across the valve with decreasing flow capacity but also reduces the effects of packing box friction on actuator positioning accuracy (dead band). As a matter of fact, you get a double effect: you get an increase in actuator force over the constant packing friction, but you also divide this friction over the larger signal span!

Again, using the above example: (a) friction, as seen by the actuator, is reduced to 78%, and (b) lower friction is distributed over 100/78% signal span. Total dead band (as caused by valve friction only) reduction is therefore 39%, $[100 \times (1 - 0.78^2)]$.

While the above schema has beneficial effects concerning actuator forces and dead band, we should not expect improvements in the "installed" valve gain. Granted, the gain change (see Figure 7-1) is now spread over a greater signal range; however, the percent of flow at which the "limit in gain change" is exceeded is still about the same.

The above example is conservative since it assumes an equal percentage plug characteristic. If the case had been a linear valve, then with the same C_v decrease, the plug travel would have to be reduced to 40% of the original travel, yielding an actuator force increase of 250% and a packing friction reduction of 84%! Simply reducing the actuator travel of a conventional valve by merely recalibrating the valve positioner will not do the same job. You still get the same actuator force and the same dead band, except the dead band is now divided into a larger signal span, thereby resulting in a lower percentage of signal!

The travel of a valve plug or rotary closure member cannot be reduced indiscriminately. For example, with some rotary valves, the seating friction near the point of shutoff becomes very high. Also, most parabolic plugs don't extend their contoured portion all the way to the seating bevel but are cylindrical for measuring purposes. Therefore, a reduction of valve travel to less than 40% of the original travel does not seem advisable. Also, such travel adjustment is not practical with valves that have an inherent low rangeability, i.e., a low ($<$20:1) ratio between maximum C_v and minimum C_v.

References

1. ISA-S75.01-1985, Flow Equations for Sizing Control Valves, Research Triangle Park, NC: Instrument Society of America (1985)

2. Baumann, H. D., "The Introduction of a Critical Flow Factor for Valve Sizing," *ISA Transactions*, Vol. 2, pp 107–111 (1963)
3. Driskell, L. R., "New Approach to Control Valve Sizing," *Hydrocarbon Processing*, pp 111–114, July (1969)
4. Bulletin No. SM-6, "Control Valve Sizing," Portsmouth, NH: H. D. Baumann Assoc., Ltd. (1991)
5. Baumann, H. D., "A Case for Butterfly Valves in Throttling Applications," *Instruments & Control Systems*, May (1979)
6. Baumann, H. D., "Effect of Pipe Reducers on Control Valve Capacity," *Instrument & Control Systems*, pp 99–102, Dec. (1968)
7. Baumann, H. D., "Viscosity Flow Correction for Small Control Valve Trim," ISA PAPER 90-618, Research Triangle Park, NC: Instrument Society of America, October (1990)

6

WHAT ABOUT FAIL-SAFE?

One of the important benefits of spring-diaphragm-actuated control valves is their ability to "fail safe" by the action of the precompressed actuator spring, i.e., to close or open the valve on air failure or, more typically, breakage of the signal wire.

Cylinder-actuated valves do not fail safe without the additions of springs or stored air pressure in a separated cylinder kept there by a pressure-actuated lockup valve. Sometimes, the flow direction of a single-seated globe valve is reversed so that the fluid forces help to close the valve plug (if fail-close action is desired).

Typically, electric actuators do not provide fail-safe action unless a standby power source, such as a relay switch-actuated battery pack, is provided. Here, of course, the periodic maintenance and recharging of the battery is a must.

The fail-safe direction (open or closed) is usually dictated by safety concerns. Typically,* "fail-close" is specified if the fluid to be handled is flammable, such as natural gas going to a burner, or is

*75% to 80% of all applications.

a heating medium going to a heat exchanger where adding too much heat to a kettle can cause boiling or can cause an exothermic reaction of a chemical. Most valves handling chemicals are specified as ''fail-close'' valves.

However, in some heating applications fail-open action is more advantageous. As an example, consider the steam tracing of a crude oil pipe. Here, too little viscosity is probably better than having the fluid come to a near standstill, particularly in the middle of winter. Another fail-open application is on cooling fluid to a motor bearing where the worst effect may be a decrease in efficiency.

Ask yourself the question, ''What would happen if this valve suddenly opens (or closes) upon failure of the air compressor or upon signal failure (computer breakdown, wire breakage, lightning strike)?'' When in doubt, check with your process people. Another question to keep in mind is, ''Does the downstream safety valve have sufficient flow capacity to handle the flow of a pressure-reducing valve when it fails wide open?'' In the same vein, consider the effects on a trapped liquid experiencing thermal expansion when a valve that is supplying a heating medium fails in the open position.

In some cases you wish the valve would remain in the last travel position, particularly if you are dealing with a short power failure to the control system. Here, the standard electric gear-type actuator is ideal. Spring-diaphragm and other pneumatic actuators have to use lockup valves to trap air pressure in the actuator if the ''fail-in-place'' mode is desired.

To guard against compressed air failure, have the pneumatic signal to the actuator pass through a valve that is kept open by the pressure in a separate air supply line. As soon as this supply pressure diminishes, it will close the lockup valve and trap the actuator pressure. To guard against signal failure, here a two-way solenoid valve can take the place of the pneumatic lockup valve and trap the air upon electrical power failure.

For added safety, you may want to consider the flow direction through the valve to guard against the consequences of valve stem failure. Here, for example, you want to specify the direction of fluid flow as ''flow-to-open'' in case the valve should fail open.

When considering the safety aspects of your control valve application, evaluate the following possible consequences in order of increasing severity:

- Decrease in product yield
- Shutdown of process
- Damage to process equipment
- Injury to personnel or loss of life

7

WHY MOST PEOPLE CHOOSE "EQUAL PERCENTAGE" AS A FLOW CHARACTERISTIC

One of the most vexing questions asked is, "What inherent flow characteristic should my valve have?" First, a definition: The "inherent flow characteristic" is the relationship between the flow coefficient C_v and valve travel. If it is linear, then you get (theoretically) the same amount of C_v increase for an equal increase in travel. Typical values are a 9.8% increase in C_v for every 10% increase in travel. Aha! You wonder why not 10% C_v? Well, at, say, 1% travel, the flow is already 2% of maximum C_v due to clearance flow between the plug and seat ring. Hence, the change is only in 9.8% per 10% increments (see Figure 7-1).

For "equal percentage characteristic," the typical C_v increase is about 43% above the previous number for each 10% of travel; for example, at 20% travel the C_v number is 143% of that at 10% travel, e.g., the C_v at 20% travel is 35; then at 30% travel it is 50, and at 40% travel, 71.6. This means that the flow increases exponentially instead of linearly. This characteristic is typically presented by a straight line on semilogarithmic paper.

Finally, there is the "quick-open" characteristic, which means the valve plug is no longer contoured but is usually a flat disk. Here, the C_v increase is approximately linear, up to a travel $\frac{1}{4}$ of the orifice

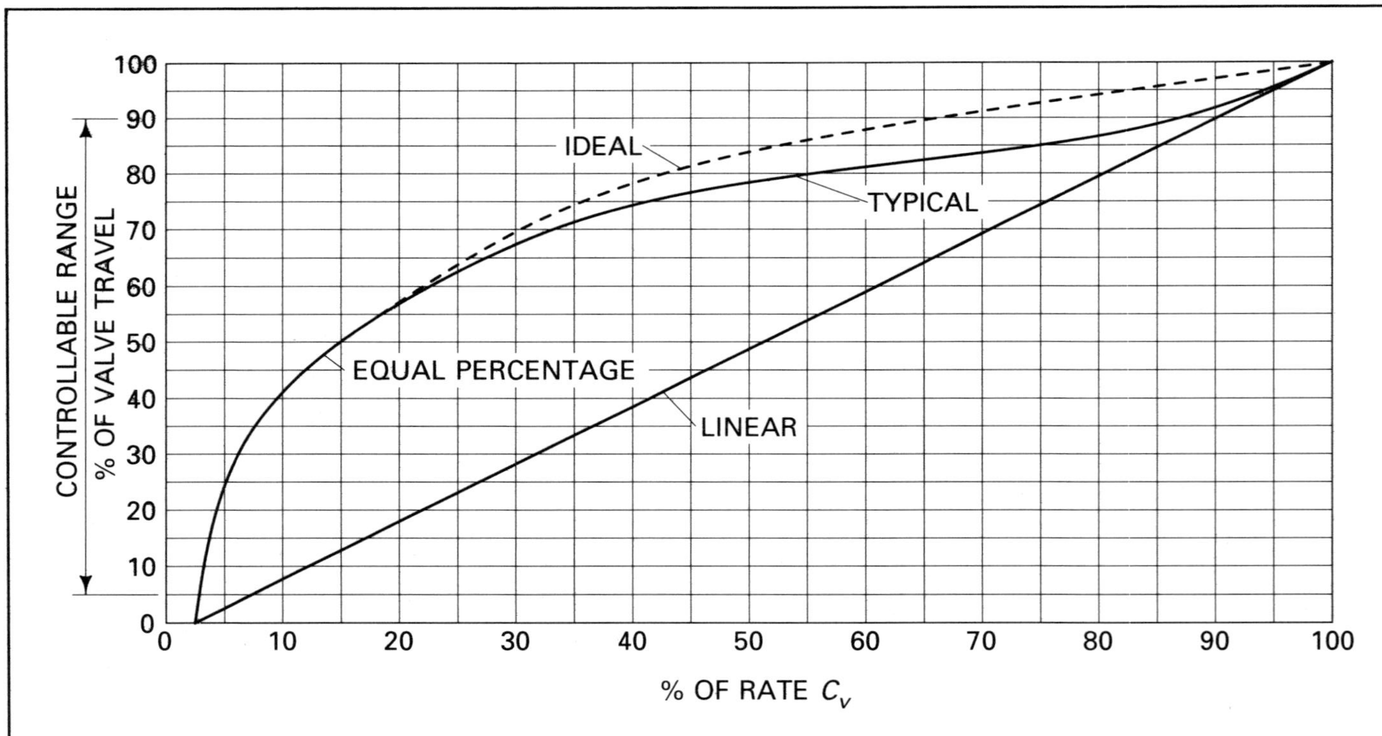

Figure 7-1. Typical inherent flow characteristics for control valves.

diameter, and then the flow stops increasing for all practical purposes. You may safely ignore this characteristic for throttling control valves.

Now a word of caution. The above flow characteristics only apply when:

- the pressure drop remains constant with flow changes,
- there will be no onset of flashing, cavitation, or sonic flow at certain flow conditions (quite possibly due to changes in the F_L factor of the valve with changes in valve travel), and
- the valve is not installed between reducers.

As you can readily see from the above, ideal conditions that prevent any of the above from occurring will prevail only in the manufacturer's flow lab. So, to account for the real world, we have to deal with the "installed flow characteristic," which is the relationship of flowing quantity to valve travel, accounting for the difference in pressure drop between maximum and minimum flow; the possibility of choked flow along the line; and the distortion that the reducer head losses create in the inherent valve flow characteristic. This installed characteristic[1] is what finally affects your controller stability.

An ideal controller PID* setting will demand a constant "gain" of the final control element. This gain is the rate of change in flow (q) for a given change of signal. Usually, the valve travel (h) is proportional to the signal. This makes $\delta(q)/\delta(h)$ = constant, for the ideal valve.

Ideally then, what we want is a "linear" installed flow characteristic, which is about a 10% increase in flow (not C_v) for every 10% increase in travel. However, nothing in life is ideal, and we, therefore, have to compromise. As a rule of thumb, the valve gain can vary ±50% from the constant gain value (does not need to be 1.0) without requiring an adjustment in the controller's PID settings. This means, for example, that one can accept an installed characteristic where the flow can vary 5% per 10% travel at one end and 15% per 10% travel on the other. This certainly makes life easier. Figure 7-2 shows an example of a typical installed valve flow characteristic. Here,

*Proportional, integral, and derivative action.

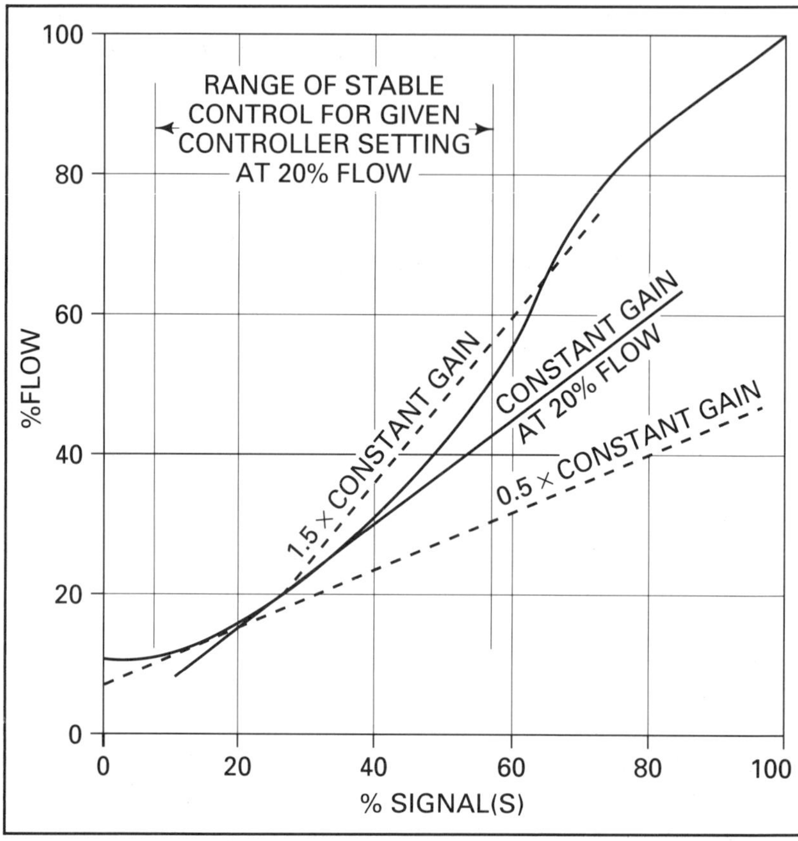

Figure 7-2. Typical installed flow characteristic of a control valve with super-imposed ±50% control valve gain variations from original controller setting, indicating limits of expected stable controller performance. Retuning of controller may be necessary if flow range exceeds those shown limits.

the controller was tuned at 20% flow or 28% of signal. The slope of the curve at that point (15% flow per 20% signal) serves as a reference. In this case, it just so happens that the gain changes in either direction. Therefore, the ±50% (i.e., 1.5 and 0.5) "constant gain" slopes determine the limits of controller stability, which are 10% flow (8% signal) and 57% flow (58% signal). Based on this information, we can now make the following rules:

First, check your process and find out what the ΔP across the valve is at the minimum required flow rate and what ΔP is at the maximum flow rate. Establish a ratio between the two. Second, if the ratio is less than 2 : 1, choose a linear inherent valve characteristic. If the ratio is above 2 : 1, choose an equal percentage plug.

So much for science. Now let us look at the more mundane aspects of the subject. As I have already stated, most valves are oversized, mostly due to overestimating pipe pressure losses that tend to lower the sizing ΔP and by allowing for future increases in throughput. Typically, you wind up with using only 40% of the valve's rated C_v. A linear valve will be about 40% open (using only 40% of the controller's signal span). An equal percentage valve, on the other hand, will be at about 75% travel. Now, which one do you think will make the better optical impression on your supervisor? You have guessed it—the valve with the equal percentage plug!

Let's consider another argument. Assume the liquid flow through a heat exchanger may vary 3 : 1 and the corresponding ΔP across the valve typically to the square of the flow rate, i.e., nine times the ΔP at $\frac{1}{3}$ the flow. Our required installed rangeability, i.e., ratio of maximum to minimum C_v, is now: $[1/(1)^{1/2}]/[0.333/(9)^{1/2}] = 1/0.111$ or 9 : 1. Dividing the 40% of rated C_v at maximum flow (see above) by nine now yields 4.4% of rated C_v at minimum flow.

For a valve with a linear characteristic, this occurs at about 3% travel or in the "danger" zone (below 5% travel) where the plug may bump against the seat following a slight controller instability. The equal percentage valve, on the other hand, has about 14% travel at the minimum flow! These are arguments you should keep well in mind.

With the availability of self-tuning controllers and computer chips that can vary the shape of the output signal curve to your valve, the question of what to choose as an inherent valve flow characteristic becomes more irrelevant. Keep in mind that most rotary control valves have "natural" inherent characteristics that cannot be defined in either of the classic definitions, yet they all seem to work well.

This all proves how forgiving most control loops are. Remember, any flow characteristic is linear when viewed under a magnifying glass. For continual processes where the flow rate does not vary by more than 20%, the gain change is less than 2 : 1, regardless of the type of inherent characteristic, so you might as well choose equal percentage due to the advantages outlined above.

A final word regarding the use of characterized cams in positioners attached to rotary control valves. These are of limited use. First, if the valve has a nearly inherent characteristic, then 4% C_v is still only 3% travel, regardless of the 14% signal input required by the equal percentage cam. Second, the modified relationship of travel to signal is correct only under static conditions, i.e., it works only with large process time constants such as with temperature control. Finally, characterized cams are inaccurate and difficult to adjust.

If the process is fast, you have a control valve that has a variable time constant with travel, which can raise Cain with the stability of the loop. Why? Because if you have an equal percentage cam to convert (for example, a linear inherent valve characteristic), it requires the valve actuator to move about 21% of its rated travel for a signal change from 90% to 80% but only 1.6% of travel to go from 20% to 10% of signal. Expressed differently, it takes 13 times longer for the valve to move at the top of the signal change than near the bottom of the signal span! Therefore, avoid cams for pressure or flow control (fast loops).

HOW TO CALCULATE THE "INSTALLED" FLOW CHARACTERISTIC

Before you sign a purchase requisition for a given control valve, you may want to know whether your controller will remain stable in a control loop where the flow rate can vary over a wide range using the maximum control valve gain change of $\pm 50\%$ mentioned earlier as a criterion. However, this cannot be done without calculating or plotting the installed flow characteristic of the valve and piping system combination, which, when completed, should look similar to the curve in Figure 7-2.

Let's assume we have a liquid flow system that consists of a storage tank, a centrifugal pump, the control valve, more than 20 feet of 6 in. pipe, four elbows, and a process vessel. In order to do the job, we need the following information.

- The maximum and minimum flow rate, given as 1250 and 250 gpm, respectively.
- The pump characteristic (converted from feet to psi), given from the manufacturer as:

gpm:	1250	1050	850	650	450	250
psi head:	110	121	131	140	146	150

- The friction loss in the pipe and elbows, calculated as 28 psi at a flow rate of 1250 gpm. Assuming there is no problem with viscosity or flashing, we can assume that the friction loss is proportional to the square of the flow rate, i.e., $\Delta P_{\text{pipe}} = 28\ \text{psi}/(\text{gpm}/1250\ \text{gpm})^2$; therefore:

gpm:	1250	1050	850	650	450	250
ΔP_{pipe}	28	19.8	12.9	7.6	3.6	1.1

- The static head change (elevation difference) from pump intake to process vessel (converted from feet to psi), given as 23 psi (remains constant).
- The backpressure in the process vessel, given as 52 psig (remains constant).

First, we find the available pressure drop across the control valve at maximum flow (1250 gpm):

$$\begin{aligned}\Delta P\ \text{valve} &= \text{pump head} - \Delta P\ \text{pipe} - \text{static head} - \text{backpressure}\\ &= 110 - 28 - 23 - 52\\ &= 7\ \text{psi}\end{aligned}$$

From the process data, we know the specific gravity = 1.1. We now can calculate the maximum required valve C_v using Equation 5-1:

$$C_v = 1250/\sqrt{7/1.1} = 496$$

Since no tight shutoff is required, and based on rangeability and cost considerations, we might consider a 6 in. angle-seating butterfly valve with a rated C_v of 825. From the manufacturer's catalog, we get the inherent flow characteristic, which is:

% of Travel:	10	20	30	40	50	60	70	80	90	100
C_v:	32	80	130	190	265	345	440	535	680	825

This tells us that the valve will be about 75% open at maximum flow—almost perfect.

Now let's find the percentage of valve travel for our other flow rates, i.e., the percentage of controller signal, which (since we did not specify a positioner cam) is about identical to present travel. Going to 1050 gpm, we again calculate the valve pressure drop as above. This is 121 psi (pump head) $-$ 19.8 psi (ΔP pipe) $-$ 23 psi $-$ 52 psi $= 26.2$ psi, and the $C_v = 1050/(26.2/1.1)^{1/2} = 215$. Extrapolating from the C_v table of the valve, we find the travel at 43.3%.

Repeating the same process for all other flow rates, we find the following:

Flow:	1250	1050	850	650	450	250
% Signal = % Travel:	75	43	31	22	15	9
Valve C_v:	496	215	136	90	57	29

The flow vs. % signal is now our installed flow characteristic as shown in Figure 7-3, curve A, which turns out to be highly nonlinear with a gain variation of about 6.7 : 1 between low flow and high flow! This obviously will result in an unstable controller situation. Unfortunately, we are stuck with the inherent characteristic of the butterfly valve. But even an equal percentage globe valve would not have been much better with the pressure drop across the valve varying over a ratio of 74 psi/7 psi = 10.5 : 1. Incidentally, if the line size had been 8 in., the required reducers would have distorted the installed characteristic even worse. Another argument for line size butterfly valves!

So, what can we do? A quick check in a popular control valve manufacturer's catalog shows the maximum C_v of a 6 in. cage valve with equal percentage trim is 357, which is too low. An 8 in. size valve is available with a rated C_v of 570. This would fit the bill, but the cost is high and it would be awkward to install a valve larger than

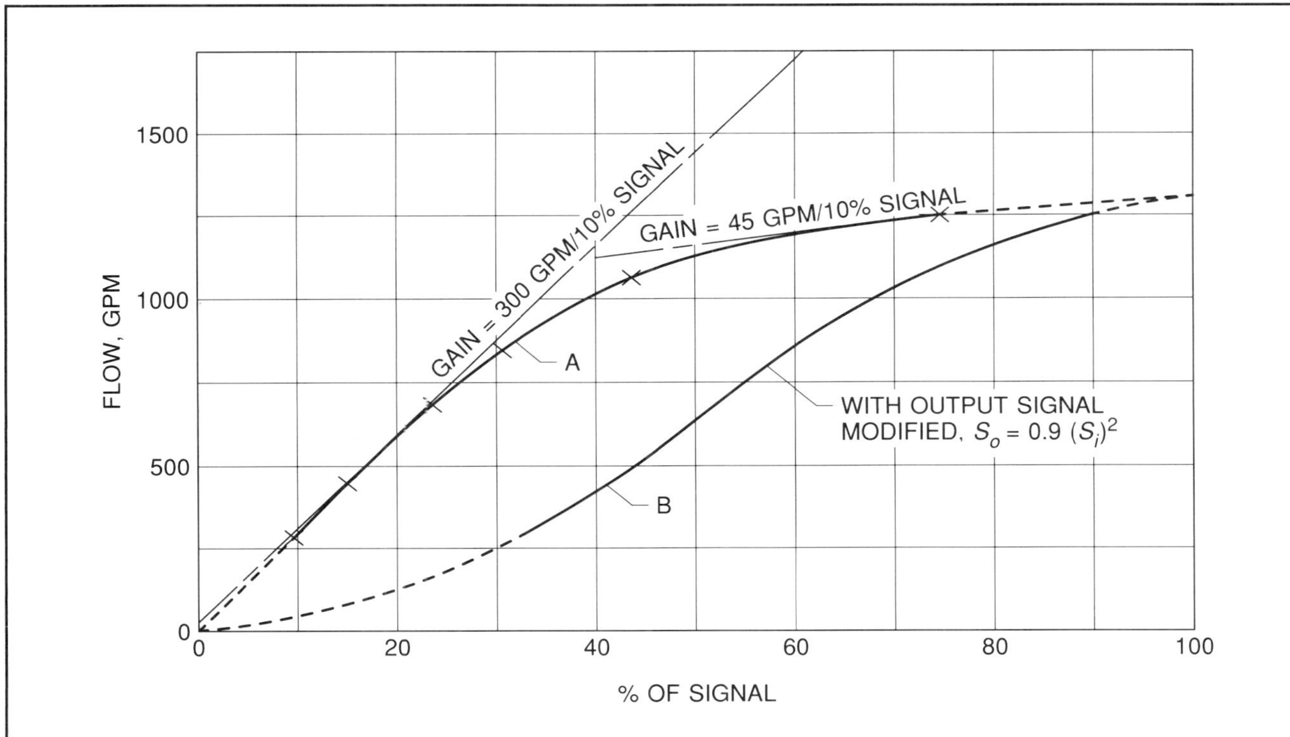

Figure 7-3. Installed characteristic (curve A) of 6 in. angle seating butterfly valve in pump and piping system causing the pressure drop to vary between 7 and 74 psi. Curve B represents the same system characteristic when internal controller logic signal of computer (S_i) is modified before being converted to analog 4-20 mA output signal S_o.

the pipe size. Besides, the characteristic would not have made that much of an improvement.[2]

A cam modified positioner characteristic is of no use for other than a slow temperature control system (see Chapter 8), so we have to rely on the modern miracle of electronics. Let's assume we modify the output signal (S_i) from a computer instead of sending it directly (via a digital-to-analog converter) to the valve in the standard 4-20 mA range. If S_i is the signal generated by the internal controller logic and S_o is the final output signal to the valve, we can program the following modification in this case:

$$S_o = 0.9(S_i)^2$$

In this particular case, I use 0.9 to bring the valve travel closer to the maximum signal and use the square function to linearize the shape of the installed flow characteristic. This yields the following relationship:

$S_i\%$:	10	20	30	40	50	60	70	80	90	100
$S_o\%$:	1	4	8	14	23	32	44	58	73	90

We can now replot the resultant characteristic, for example, by taking the flow at 32% travel (S_o) from curve A (880 gpm) and moving it to the corresponding 60% computer signal (S_i) point. After transferring all points, we now have the relationship between flow and controller logic signal (S_i) in the form of curve B. While we still have some gradual change in slope, the resultant gain* variations from 1.92 to 0.72 (240 gpm/10% to 90 gpm/10%) are all within the stated limitation of $\pm 50\%$. If the controller is set at a gain of 1.35, then the upper limit $= 1.35 + 50\% = 2.025$, and a lower limit $= 1.35 - 50\% = 0.68$, both above and below the actual gain variations.

*The control valve gain is the percent of flow change divided by a percentage of signal change. Thus, a gain of one, in this case, will yield 10% of flow (125 gpm) increase per each 10% of travel increase.

References

1. Baumann, H. D., "How to Assign Pressure Drop Across Control Valves for Liquid Pumping Systems," *Proceedings of the 29TH Annual Symposium on Instrumentation for the Process Industries*, Dept. of Chemical Engineering, Texas A & M University, Jan. 16, 17, 18 (1974).
2. Boger, H. W., "Flow Characteristic for Control Valve Installations," *ISA Journal*, pp 50–54 November (1966).

8

VALVE POSITIONERS

WHEN TO USE VALVE POSITIONERS

Positioners are sometimes a necessary evil—an evil, nevertheless, that should be avoided if possible. Broadly speaking, use of a valve positioner adds another control loop within a control loop where the set point is the controller signal to the positioner; the feedback is the lever attached to the actuator stem sensing the latter's position; the controller is the flapper-nozzle, or pilot valve, sensing the difference between the force created by the controller signal and that of a feedback spring attached to the lever; and the final control element is the relay that sends air pressure to the valve actuator.

Like any other loop, it has its own dead time, time constant, and gain, which quite often interfere with the original process control loop dynamics[1] and makes the task of the process controller more difficult in fast loops, i.e., for pressure or flow control.

So much for outlining the problem. There are, of course, instances in which the use of a valve positioner is unavoidable:

- When a cylinder actuator is used that requires air pressure in excess of 15 psi. Here, the positioner is used as a pressure multiplier.
- When a valve is split ranged, i.e., 3–9 or 9–15 psi signal to two valves from a common controller. The preferred way is to use two separate electronic signals from a common controller (use two I/P transducers).
- When valves (usually one of two) are operated in a reverse-acting mode, i.e., when a valve that for safety reasons fails closed (air-to-open actuator), shall be operated in the signal-to-close direction.
- When the expected dead band of the actuator/valve combination exceeds 5% of signal (0.6 psi or 0.8 mA) span due to excessive stem or trim friction.
- When a booster has to be used for high stroking speed and may have excessive dead band.

I am not aware of any other reasons why a valve positioner should be used except perhaps to increase the speed of response in case a valve is located too far away from a control room-installed I/P transducer. But even there, a valve mounted-transducer or booster relay will perform better with an ordinary valve not meeting any of the other criteria above. Think also of how much money will be saved not only in hardware but in piping, air sets, and maintenance as well!

However, if a positioner is not avoidable, try to specify a modern version that offers the ability to vary its (and that of the connected valve's) operating characteristic. Past valve positioners were like process controllers without proportional band and reset adjustments. One may get lucky and everything would be stable; then again, the valve (or the complete control loop) might ''hunt'' in an unstable mode. It is educational to walk through a typical refinery or chemical plant and count all the control valves that are in operation *with the valve positioner disconnected or in the bypass mode!*

To make life easier for today's process operators, many new positioners offer speed and gain adjustments or a combination of the two. This is like adding the adjusting knobs back onto the controller housing. This enables you to change the frequency response characteristic (or time constant) of the final control element in optimum relation to that of the process. You probably know that a loop cannot be stable if the process time constant is equal or close to that of the controller/valve combination. In order to have a stable loop, the ratio between the valve and the system time constants should be at least 3 : 1 (or 0.3 : 1)—better yet, 5 : 1 (or 0.2 : 1). Ideally, the valve/positioner loop should be faster than the process loop, which is sometimes hard to accomplish.

With a conventional valve/positioner combination, you had no choice but to either reduce the controller settings (high proportional band—long reset action) or crimp the output tubing from the positioner to the actuator in order to slow the valve down, particularly when the positioner was unstable due to excessive valve stem friction. Now, hopefully avoiding the pitfalls of having somebody ignorant playing with the adjustments in the field, you are able to change the speed of the valve travel (typically by 5 : 1) and/or the open-loop gain* setting (sensitivity) of the positioner (typically by 10 : 1) even while the valve is in operation. (A word of caution: you may have to realign the zero adjustment somewhat after changing the gain setting.)

A typical example of how the time constant of the valve can be varied is shown in Figure 8-1. With maximum air flow and a lower gain setting, the "break frequency" is 0.75 cps. At the other extreme, an air flow of only 0.6 m^3/h and a higher gain of 67 (equivalent to 1.5% proportional band) produces a break frequency of 0.32 cps—almost a 220% change in the time constant. Even if you don't have the faintest notion about your process dynamics, keep these guidelines in mind:

- If you have a slow process such as temperature control, set the gain at or near maximum. This way the valve will respond to the smaller incremental signal change (due to reset action

*The definition of the "open-loop gain" of a positioner is the change in output signal over the change in input signal (when the feedback is disconnected). Thus, a gain of 200 : 1 causes a 20 psi output signal change per 0.1 psi input signal change. Gain is the reciprocal of percent band; hence, a gain of 200 : 1 is equivalent to 0.5% proportional band. Some positioners have gains as low as 30 : 1; however, some exceed 200 : 1.

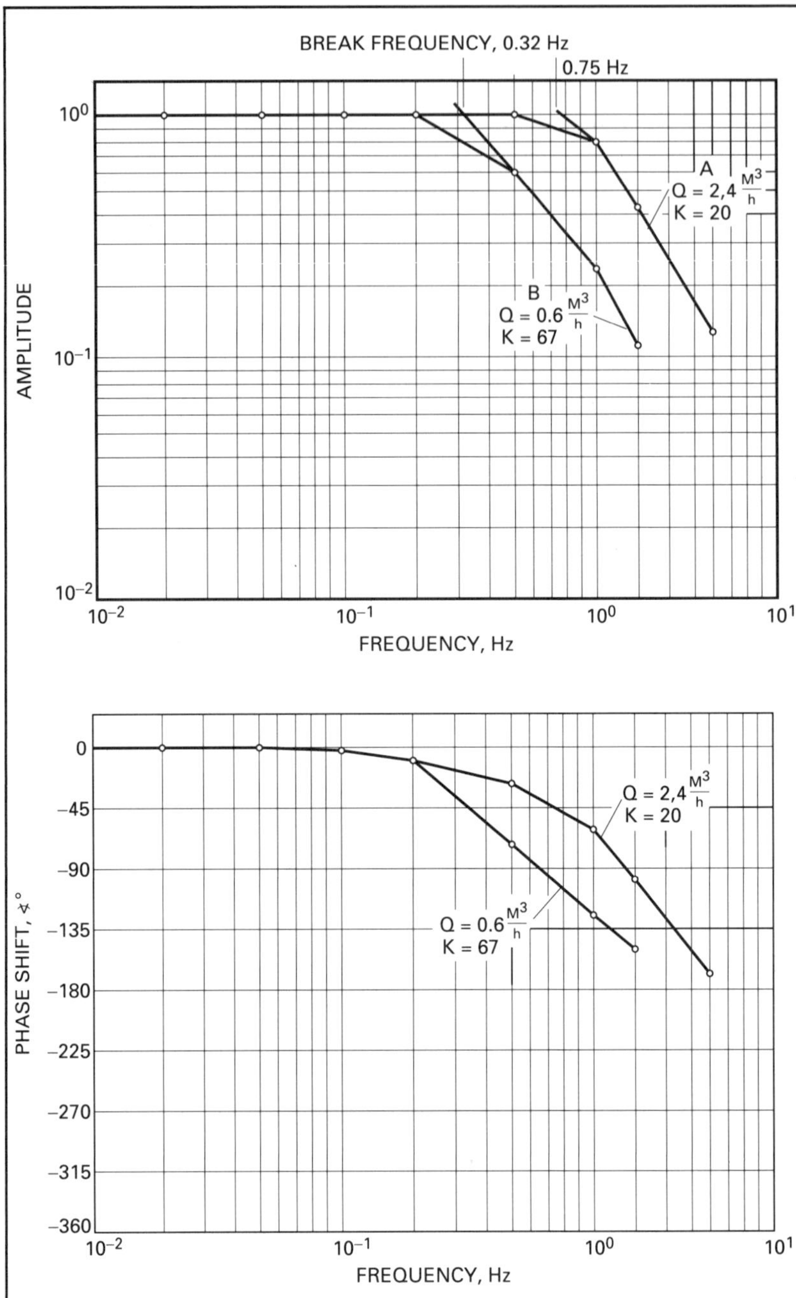

Figure 8-1. Typical adjustable range of dynamic performance of actuator/positioner combination with valve positioners having separate gain and speed adjustments. (Courtesy of SAMSON AG).

of the controller). However, set the speed (air output) near minimum, thus avoiding an unstable actuator.

- If you have a fast process such as pressure or flow control, set the gain near minimum and set the actuator travel speed at its maximum value, thus creating a very responsive but stable valve.

WHAT TO LOOK FOR WHEN SPECIFYING POSITIONERS OR TRANSDUCERS

About 90% of all control loops use a 4–20 mA signal, so you are forced to transduce the electronic signal to a pneumatic one. If you decided to employ a positioner, then an I/P positioner is your choice; otherwise, an I/P transducer has to be specified.

First, consider whether or not the valve is installed in an area where flammable liquids are handled. You may need an explosion-proof (or intrinsically safe) version. Second, check the vibration level at the valve location. For example, if the valve is close to a centrifugal pump, pipe vibrations can exist in the 60 Hz, or multiple modes thereof, frequency range. Any good I/P positioner or transducer should maintain steady output under vibration levels of at least 3 g's at 30 cps and 1 g at 100 cps (g = acceleration of gravity, 31.1 ft/s^2). Third, consider the quality of air supply. Do you use dehydrated and oil-free instrument air, or air taken directly off the factory compressor? In the last case, avoid all positioners that utilize built-in amplifying relays. The small passages may be subject to plugging. Fourth, consider the effects of weather, if an outdoor installation, or high ambient temperature, if the valve is located near a furnace, for example. Finally, where good frequency response is a must (flow control, turbine or compressor bypass, pressure control, etc.), do not use a transducer that is rack-mounted on a remote instrument panel. Locate the I/P transducer on the valve actuator to avoid signal delays.* Make

---

*This is especially important for "fast loops" such as flow or pressure control applications, since time lags due to long transmission lines between transducers and valve actuators may vary from 0.2 to 2 seconds per 10% signal change.[2]

sure the air output capacity of the transducer or positioner is sufficient to give adequate travel speed. Note: Some rack-mounted I/P transducers have air capacities of 1–4 scfm, which limits their use to temperature applications and to the operation of smaller valves only.

As a rule of thumb, allow at least one scfh at 20 psi supply for each in.^{2} of diaphragm area (3–15 psi signal). For example, a valve with a 100 in.^{2} diaphragm actuator should need a positioner with at least a 100 scfh air output capacity. For very large actuators, you may need a booster relay between instrument and actuator.

Other desirable features are: all exterior linkages to be type 18-8 stainless steel; all aluminum parts to be anodized (interior and exterior) *and* epoxy coated; positioners to have at least a built-in gain adjustment; I/P transducers to have the capability for 1–17 psi air output range (see Chapter 9 on off-balance plug forces); and, if of "closed-loop" design (with built-in pressure transducer feedback), should have a gain or dead band adjustment to avoid instability when mounted close to a small diaphragm actuator.

References

1. Lloyd, S. G., "Guidelines for the Application of Valve Positioners," *Proceedings of the Texas A&M Instrumentation Symposium*, January (1969).
2. St. Clair, D. W., *Controller Tuning and Control Loop Performance*, p. 19, Newark, DE: Straight Line Control Company.

9

THE MYSTERY OF LINE PRESSURE-PRODUCED VALVE STEM FORCES, OR SELECTING THE CORRECT ACTUATOR SIZE

The selection of an adequate valve actuator is usually left up to the vendor. Most of the time this works well, especially if the pressure conditions the valve has to handle are low or the valve size is small. However, you may want to know some of the consequences in selecting the wrong (usually too small) actuator size. Remember, the valve vendor who got the job was probably the lowest bidder and, quite often, the bidder who cut the size of actuators in order to reduce costs. What follows will tell you what can happen after installation.

THE VALVE DOES NOT CLOSE PROPERLY

The economically dictated demise of bypass installations around a control valve has led to the requirement that a control valve have ANSI/FCI Class IV seat tightness (to double as a shutoff valve when the system is down). This works well as long as the initial spring setting, the air pressure, and size of the actuator are adequate to overcome the fluid pressure acting on the valve plug (assuming a single-seated globe valve).

To start with, it is important to tell the vendor what the maximum inlet pressure is! This is usually the pressure differential across the plug when the valve is shut off and *not* the lower inlet pressure and substantially higher outlet pressure that the valve sees during normal throttling or that is used for calculating the valve C_v. A responsible valve vendor will always assume the valve has to shut off against the maximum inlet pressure and with a zero gage outlet pressure. Therefore, P_1 has to be stated on the specification form (see Figure 9-1). For example, a heating valve with a 2 in. orifice handling hot oil will see, at the normal flow rate and 75% valve travel, an inlet pressure of 125 psig and a downstream pressure of 110 psig. The stem forces sensed by the actuator under these conditions are $(2^2\pi/4) \times (125 - 110) = 47$ lbs.* However, if called to close, the inlet pressure will be 140 psig due to the increase in pressure following the pump's characteristic, while the downstream pressure may be zero. In this case, the minimum stem force will be 140 psi times the seating area of the plug (use for the seating diameter at least $\frac{1}{16}$ in. plus the orifice diameter) plus the packing friction. Hence, stem force (closed) = $(2.063^2\pi/4) \times 140$ + packing friction.

Now packing friction can vary considerably; for example, tests with TFE containing packing material showed that starting friction (or breakaway friction) was 1.5 times that of dynamic (constant movement) friction.[1] However, as a rule of thumb, use 1.5 × stem area × max P_1. In this case, assume a $\frac{3}{8}$ in. stem, PF = $1.5(0.375^2\pi/4) \times 140$ psig = 23 lbs. Total stem force (closed) is, therefore, 468 + 23 = 491 lbs.

As you can see, there is more than a tenfold increase in actuator force requirement. In case of a piston actuator, and assuming the availability of at least 80 psi air, the minimum cylinder area should be 491 lbs/80 lbs/in.^{2} = 6.2 in.^{2} or a 3 in. piston diameter.

If you want a spring-diaphragm actuator to fail closed and you specify a valve positioner, a 6-15 psi span spring inside a 100 in.^{2} actuator would be adequate. Here, the output signal from the positioner would vary as follows:

---

*Neglecting the area of the valve stem for simplicity.

©ISA S20.50, Rev. 1 CONTROL VALVE DATA SHEET Second Printing

ISA®	PROJECT ____	DATA SHEET ____ of ____
	UNIT ____	SPEC ____
	P.O. ____	TAG ____
	ITEM ____	DWG ____
	CONTRACT ____	SERVICE ____
	*MFR. SERIAL	

1	Fluid				Crit Press PC		
		Units	Max Flow	Norm Flow	Min Flow	Shut-Off	
2	Flow Rate					—	
3	SERVICE CONDITIONS: Inlet Pressure						
4	Outlet Pressure						
5	Inlet Temperature						
6	Spec Wt/Spec Grav/Mol WT					—	
7	Viscosity/Spec Heats Ratio					—	
8	Vapor Pressure P_v					—	
9	*Required C_v					—	
10	*Travel	%				0	
11	Allowable/*Predicted SPL	dBA	/	/	/	—	
12							

13	LINE	Pipe Line Size In ____	53	ACTUATOR	*Type ____
14		& Schedule Out ____	54		*Mfr & Model ____
15		Pipr Line Insulation	55		*Size ____ Eff Area ____
16	VALVE BODY/BONNET	*Type ____	56		On/Off ____ Modulating ____
17		*Size ____ ANSI class ____	57		Spring Action Open/Close ____
18		Max Press/Temp ____	58		*Max Allowable Pressure ____
19		*Mfr & Model ____	59		*Min Required Pressure ____
20		*Bony/Bonnet Matl ____	60		Available Air Supply Pressure: ____
21		*Liner Material/ID ____	61		Max ____ Min ____
22		End In ____	62		*Bench Range ____ / ____
23		Connection Out ____	63		Actuator Orientation ____
24		Flg Face/Matl ____	64		Handwheel Type ____
25		End Ext/Matl ____	65		Air Failure Valve ____ Set at ____
26		*Flow Direction ____	66		
27		*Type of Bonnet ____	67		Input Signal
28		Lub & Iso Valve ____ Lube ____	68	POSITIONER	*Type ____
29		*Packing Material ____	69		*Mfr & Model ____
30		*Packing Type ____	70		*On Incr Signal Output Incr/Decr ____
31			71		Gauges ____
32	TRIM	*Type ____	72		*Cam Characteristic ____
33		*Size ____ Rated Travel ____	73		
34		*Characteristic ____	74	SWITCHES	Type ____ Quantity ____
35		*Balanced/Unbalanced ____	75		*Mft & Model ____
36		*Rated C_V ____ F_L ____ X_T ____	76		Contacts/Rating ____
37		*Plug/Ball/Disc Material ____	77		Actuation Points ____
38		*Seat Material ____	78		
39		*Cage/Guide Material ____	79	AIR SET	*Mfr & Model ____
40		*Stem Material ____	80		*Set Pressure ____
41		____	81		Filter ____ Gauge ____
42			82		
43	SPECIALS/ACCESSORIES	NEC Class ____ Group ____ Div ____	83	TESTS	*Hydro Pressure ____
44		____	84		ANSI/FCI Leakage Class ____
45		____	85		____
46		____	86		
47		____			
48		____			
49		____			
50		____			
51		____			
52		____			

Rev	Date	Revision	Orig	App

*Information supplied by manufacturer unless already specified.

ISA FORM S20.50, Rev. 1

Figure 9-1. Control valve data sheet from ISA-S20.50-1983. (By permission of the Instrument Society of America)

- At zero travel (valve shut), the force created by the spring is 6 psi × 100 in.^{2} = 600 lbs. However, since the seating stem force here is 491 lbs (trying to push up), it leaves only 600 − 491 lbs = 109 lbs for the actuator to open the valve. Hence, the air signal is 109 lbs/100 in.^{2} = 1.1 psi!
- Remember, at 75% travel (normal operating conditions) our stem forces are reduced to 47 lbs. At this point, our spring exerts a force of [(15 psi − 6 psi) × 0.75 + 6 psi] × 100 in.^{2} = 12.75 psi × 100 in.^{2} = 1275 lbs. Subtracting from this value the 47 lbs acting up (i.e., against the spring) = 1275 − 47 = 1228 lbs net spring force. This translates into a positioner output signal of 1228 lbs/100 in.^{2} = 12.28 psi.
- We thus have an installed positioner output signal span of 1.1 to 12.28 psi between zero and maximum flow at 75% travel. This is in contrast to a controller signal *to* the positioner of 3–12 psi. This contradicts a common fallacy that the output signal of a positioner has to match the input signal!

As you can see from the above example, if you were to disconnect the positioner, the valve would never close with a constant 3 psi input signal from an I/P transducer (electro-pneumatic converting instrument). This is exactly what has happened in many plants that utilize computers or single-loop digital controllers for process control.[2] These devices lock in the 4 mA signal at zero process conditions.* If you have an I/P transducer that is calibrated to have a 3–15 psi output range for a 4–20 mA input signal (and most are), you will continuously feed 3 psi to the valve and, as a result, could cause valve leakage. Unfortunately, this situation most often leads to a call to the vendor informing them that their valve is bad (lousy quality, etc.).

What can be done about it? Well, in order of cost effectiveness and engineering preference:

- If you can, reduce the zero setting on your I/P transducer to 1 psi at 4 mA, or specify an I/P transducer with a 1–17 psi output per ANSI/FCI Standard 87-2.

---

*It is possible to adjust your computer to under- or overrange the signal by 5% or 0.8 mA. However, allowing the signal to decrease below 2 mA can cause failure of certain I/P transducers, with dire consequences for your process loop!

- Ask your valve vendor to select an actuator spring with an 8–15 psi "bench range." (This, in the above example, will give you an installed range of between 8 minus 4.9,* when seated, and 15 minus 0.5** psi, at 100% travel, which is 3.1 to 14.5 psi, close enough to be called 3–15 psi range!)
- Specify a valve positioner.

Please note that simply cranking up the initial spring setting from, say, 3 psi to 6 psi won't do. You will wind up with a bench range of 6–18 psi and your valve will not open enough to handle flow upsets. (Remember, a standard I/P signal cannot exceed 15 psi at 20 mA input!)

THE VALVE IS UNSTABLE

The instability caused by dynamic fluid forces on a plug, disk, or ball should not be confused with positioner or control loop instability (sometimes hard to distinguish). As was shown in the prior example, forces acting on a valve plug or across a butterfly valve (or a rotating plug, for that matter) can vary drastically with flow conditions. Many papers have been written on the subject,[3] most of them requiring you to become a math major! Therefore, I will try to keep the matter simple.

For a valve trim to be stable, the change in stem force per inch of stem motion has to be less than the change in stem force per inch of travel exerted by the actuator. The latter is called the actuator stiffness factor, K_n, and is expressed in lbs/in. (also see ISA-RP 75.18-1989). For a spring-diaphragm actuator this is simply the spring rate of the actuator spring (or all of the springs combined if multi-spring), ignoring the pneumatic damping.

With piston actuators, things get a bit more complex. Here, it is the force generated by compressing the internal actuator air pressure

---

*492 lbs seating force/100 in.^{2}

**47 lbs stem force/100 in.^{2}

when the piston is moved over a given distance that determines the stiffness. Obviously, this force varies with stem position, initial cylinder volume L, and absolute air pressure. As a rule, piston actuators are much stiffer[4] than spring-diaphragm actuators and, therefore, are able to operate single-seated globe valves in the flow-to-close direction—a feat normal spring-diaphragm actuators cannot handle.

The general practice is to install single-seated globe valves in the flow-to-open direction, where the problem of dynamic instability seldom arises. However, it can be a problem in force-balanced cage valves and angle valves that handle erosive or flashing fluids where discharging occurs downstream of the valve plug (flow tending to close).

What happens here is that not only does the effective area of plug increase as you approach the seating point but, in addition, the downstream pressure decreases rapidly, roughly to the square of the flow rate decrease. We all have experienced this phenomenon in the bathtub; with water in the tub, we let go of the chain slowly and the rubber plug approaches the drain, but, at about $\frac{3}{16}$ in. from the seat, the plug suddenly surges down and slams onto the seat. The same thing happens in a valve. The trick is to have the actuator stiff enough to maintain hold of the valve plug.

As a rule of thumb, consider that the off-balance plug area goes from zero to maximum within a distance that is the lesser of either $\frac{1}{3}$ of valve travel or $\frac{1}{4}$ of the orifice diameter. Further assume that the differential pressure will increase from $P_1 - P_2$ at normal flow to P_1 maximum within this same distance.

Here is an example: A 3 in. angle valve with a 2.5 in. orifice diameter has a $\frac{1}{2}$ in. valve stem diameter. Normal $P_1 = 165$ psig; normal $P_2 = 125$ psig; maximum $P_1 = 175$ psig; flow-to-close actuator travel = 1.5 inches. Since the area of the stem is relatively small, we can ignore this and assume the off-balance plug area is the 2.5 in. orifice area, which is 4.9 in.^{2}

First consider the distance of area change: $\frac{1}{3}$ of 1.5 in. travel = 0.5 in.; $\frac{1}{4}$ of 2.5 in. orifice diameter = 0.625 in.; therefore, we chose 0.5 in., being the smaller. The rate of change in stem forces is now:

$$\text{Off-bal. area} \times \Delta P\ \text{change/Distance} \tag{9-1}$$

(where ΔP change = max. $P_1 - \Delta P$) or 175 − (165 − 125) = 135 psi. Therefore, from Equation 9-1, 4.9 in.^{2} × 135 psi/0.5 in. = 1323 lbs/in.

Assuming the actuator is spring-diaphragm with an area of 100 in.^{2} and a 3–15 psi bench range, this makes the spring rate (15–3 psi) × 100 in.^{2}/1.5 in. travel = 800 lbs/in. This is quite insufficient, and we either have to recommend a 175 in.^{2} actuator or change the bench range of the spring (or springs) to 6–30 psi. This would increase the actuator stiffness to 1600 lbs/in., which will provide a sufficient safety margin. Air compression within the actuator will increase the stiffness and, therefore, provide a safety factor, although such stiffness is much less than in cylinder actuators due to larger air volume.

Rotary control valves have similar stiffness problems. They are more complex since we have to deal with torque instead of force. Consider the combined torque characteristic of a butterfly valve (shown in Figure 9-2 from Figure 4 in Reference 5). Here, the greatest change in torque per degree occurs at 60°, where the combined torque changes at a rate of (0.011 in.-lbs/d_i^3) × P_1 for ±5° travel, i.e., from 0.064 @ 55° to 0.075 @ 65°. Let's assume we have a 10 in. valve with a maximum differential pressure (when the valve is closed) of 87 psig. We have a torque change at 60° opening (and assuming the valve has to open above 60° to pass the maximum flow) of 0.011 in.-lbs × 10 in.^{3} × 87 psi = 960 in.-lbs per 10°.

Assume we have a cylinder actuator with a 75 in.^{2} area that travels 4 in. to move the vane from 0° to 90° opening. Neglecting angular cosine effects, we can assume that at a 65° opening the piston travel will be at (4 in./90°) × 65° = 2.88 in., and at 55° vane travel = (4 in./90°) × 55°, at 2.44 inches.

What we need to know is the volume change in the actuator going from 2.44 in. to 2.88 in. piston travel and assuming 0.3 in. of overtravel (O.T.) in the cylinder. Let's calculate the ratio of air volume between both travel positions:

$$\text{Volume ratio} = \text{Area} \times (\text{Init. travel} + \text{Overtravel}) / \text{Area} \times (\text{Final travel} + \text{Overtravel})$$

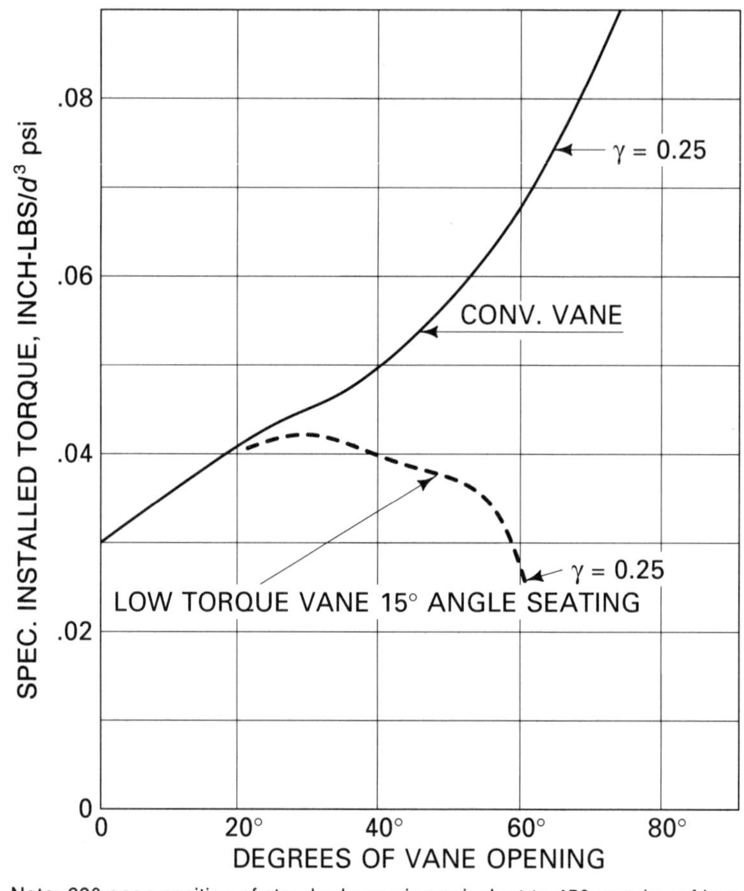

Note: 60° open position of standard vane is equivalent to 45° opening of low-torque vane due to added seating angle.

Figure 9-2. Dynamic torque characteristic of swing-through butterfly vs. angle-seating, low-torque vane.

Therefore,

$$\begin{aligned}\text{Volume ratio} &= 75\ \text{in.}^2\ (2.88\ \text{in.} + 0.3\ \text{in.})/\\ &\qquad 75\ \text{in.}^2\ (2.44\ \text{in.} + 0.3\ \text{in.})\\ &= 239\ \text{in.}^3/206\ \text{in.}^3\\ &= 1.16\end{aligned}$$

Assuming further that the positioner output pressure in the cylinder at 65° travel is 65 psia, we can now calculate the force change due to air compression if the actuator should suddenly have to move from 65° to 55° due to dynamic torque effects (remember that in most rotary valves torque tends to close).

$$\begin{aligned}\text{Act. force change} &= (\text{Volume ratio} - 1) \times \text{Area} \times \text{Air pressure}\\ &= (1.16 - 1) \times 75 \text{ in.}^2 \times 65 \text{ psia}\\ &= 780 \text{ lbs}\end{aligned} \tag{9-2}$$

All we have to do now is convert this force change to torque. Since we know the travel is 4 in. at 90° rotation, we can assume that the effective length of the lever turning the shaft is equal to actuator travel/2 × tan 45° = 4/(2 × 1) = 2.00 inches.

Thus, the force change of 780 lbs translates into a torque change of 780 lbs × 2.00 in. = 1560 in.-lbs/10° travel. This is above the dynamic vane torque change of 960 in.-lbs/10°, and we could expect stability of operation since the friction of the piston seal will provide extra resistance to sudden change.

If this had been a spring-diaphragm actuator with the same 4 in. travel, a torque requirement of 960 in.-lbs per 10° vane travel would demand a 960 in.-lbs/2 in. eff. lever = 480 lbs force change per 0.44 in. actuator travel and, therefore, would require a 480 lbs/0.44 in. = 1091 lbs/in. spring rate. Neglecting added stiffness due to air compression, this is equivalent to a 180 in.^{2} actuator with a 6–30 psi bench range. Quite a drastic increase in actuator size.

While the above example assumed a conventional vane, so-called "low torque" vanes would reduce the stiffness requirement and would, therefore, be very effective when allowing for the use of spring-diaphragm actuators despite their lower stiffness factor. Note: a 60° opening with the 15 in. angle-seating, low torque vane occurs at 45° vane travel (see Figure 9-2).

The following is a summary of the stiffness equations.

- *Spring-Diaphragm Actuator* (ignoring lesser stiffness due to air compression*)

---

*Assumed to be isothermal.

$$K_n = (\text{Diaphragm area} \times \text{signal span})/\text{Travel} \qquad \text{lbs/in.}$$

- *Cylinder Actuator* (single-acting)

$$K_n = \{[A_c(h_i + h_o + \Delta h)/A_c(h_i + h_o)] - 1\} \times S \times A_c/\Delta h \qquad \text{lbs/in.}$$

where:

A_c = effective area of cylinder, in.^{2}
h_i = initial travel position, in.
h_o = overtravel of piston (travel not used), in.
S = signal pressure at given travel position, psia
Δh = change in travel due to disturbance, in.

- *Cylinder Actuator* (double-acting). Results from single-acting cylinder actuator times two.

References

1. Jeschke, N., et al., "Untersuchungen der Stopfbuchsreibung an Stellventilen mit pneumatischen Membranantrieb," *RTP*, Volume 3, West Germany (1968).
2. Baumann, H. D., "The Need for Pneumatic Power Signals for Control Valves," *INTECH*, pp 34–41, Jan. (1988).
3. Keith, G. A., "Analytical Prediction of Valve Stability," ISA Paper 838-70, Research Triangle Park, NC: Instrument Society of America (1970).
4. Schafbuch, P. J., "Fluid Inertia Effects on Unbalanced Valve Stability," ISA Paper 85-0207, Research Triangle Park, NC: Instrument Society of America, (1983).
5. Baumann, H. D., "A Case for Butterfly Valves in Throttling Applications," *Instrument and Control Systems*, pp 35–38, May (1979).

10

HOW TO INSTALL A CONTROL VALVE

Having properly selected a given control valve does not automatically ensure that it will do the right job. Proper installation is almost as important. Unfortunately, this might depend greatly on the experience of the person who designed the piping layout. However, it may not hurt to ask a few questions regarding the planned control valve installation. If nothing else, it shows him or her that you care. On the other hand, you might just prevent a major mistake you may be blamed for latcr.

Here, then, are some, but by no means all, of the important considerations regarding control valve installations.

Accessibility. Is there sufficient space to remove the actuator when repair is necessary? Can the flange bolt or the tie rods be removed? (This can be a problem, especially with flangeless valves, if the body is installed between pipe reducers.) Can the positioner gages or the valve travel indicator be seen? Can the positioner be adjusted?

Considerations for Manual Operation. In many critical applications you may want to consider a manually operated throttling valve in a bypass around the valve, with appropriate block and drain valves on either side of the control valve that allow for the repair or replacement of the latter without shutting down the process (ascertain if it is possible to run the plant manually).

Location. Is the location of the valve chosen to permit repair? Is the location optimized to help prevent cavitation or, what is worse, permit cavitation to happen in a normally flashing downstream environment because the outlet pipe is too long? If the valve is on the bottom of a steam header, place a steam trap close to the valve inlet. Do not place an elbow or pipe tee less than six pipe diameters downstream of a butterfly or a ball valve to avoid interference with the valve's flow capacity.

In high pressure gas or steam pressure-reducing applications, try to have only straight pipes downstream of the valve or use long sweeping elbows to avoid additional pipe noise. Allow for drainage through the valve. This is especially important for sanitary valves and for valves that handle dangerous fluids.

Consider the possibility of corrosion caused by leaky flange gaskets or valve stem packings on equipment located *below* the valve. Is the valve located near a source of extreme heat or where the accessories might freeze up in winter? You may have to consider the ambient temperature rating of the actuator or positioner, or heat trace the valve or air lines. Does the valve have to be fire-safe or must the electrical accessories be explosion-proof due to location in an area considered hazardous?

Other Things That Can Go Wrong

- The installer forgets to pre-torque the tie rods that hold a flangeless control valve, especially when it will control a cold fluid.
- The installer paid no attention to the flow arrow on the valve body.
- No filter was placed ahead of a valve that has a close clearance valve plug ($C_v < 0.1$).
- The valve body of a butterfly or ball valve was misaligned with the piping flanges. This can reduce the maximum valve C_v by up to 20%!
- No shutoff valve was provided in the pneumatic supply line to the positioner.

- An elastomer-lined butterfly valve is installed between slip-on type line flanges, causing the liner to fail because it is not properly retained due to the narrow width of the gasket surface.
- A PTFE-lined corrosion-resistant valve is installed between plastic-lined raised-face line flanges, causing the PTFE valve surface to "cold flow" and cause leakage between the body and pipe flanges.
- A bleed or drain valve was not installed in the pipe to relieve the pressure in the line when the control valve is shut down.
- The installer insulated the extension bonnet on a cryogenic valve, causing the packing box to freeze up.
- No protective metal shroud was placed around the tie rods and wafer-style valve body in an area where there is a potential fire hazard.
- The bellows seal (sealing diaphragm) was not removed from a valve during hydrostatic testing of the valve and piping system, causing bellows (diaphragm) rupture.

Of course, your primary concern should always be safety. The installation certainly should meet all safety code requirements and should be investigated to eliminate all hazards, including the possibility of causing pollution. Then, consider saving your company a bundle of money by ensuring that this valve will last for at least 20 years *without* being the cause for a very expensive shutdown of the plant or process.

References

1. API RP550, 2nd Edit., Manual on Installation of Refinery Instruments and Control Systems, New York, NY: American Petroleum Institute.
2. *Valves, Piping and Pipe Line Handbook*, 2nd Edit., Surrey, UK: The Trade and Technical Press Limited (1986).
3. *ISA Handbook of Control Valves*, 2nd Edit., Research Triangle Park, NC: Instrument Society of America (1976).

11

WHEN DO I NEED TO "HARD FACE" THE VALVE TRIM AND OTHER QUESTIONS CONCERNING VALVE MATERIAL

Besides corrosion, which can be handled by consulting a good handbook[1] when selecting a suitable material for wetted parts, erosion is the cause of most valve failures. We need to distinguish the causes of erosion in order to effect a durable solution.

EROSION CAUSED BY SOLID PARTICLES IN THE FLUID STREAM OR GRANULAR MATERIALS SUCH AS COAL SLURRY

If the velocity of the erosive fluid is low enough (say, less than a 10 psi valve pressure drop), then elastomer lining works very well, particularly with natural rubber or GRS (BUNA-S) types of elastomers, and a pinch valve or rubber-lined butterfly valve should be considered. Remember, you need a pressure drop of at least one psi to keep particles in suspension (3 to 7 feet/second velocity). For higher pressure drop or where temperature is elevated, consider angle valves in the "flow tending-to-close the plug" direction. Try to discharge the flow into a downstream pipe that has a diameter of at least four times the orifice size in order to avoid cutting through the pipe wall. Hard facing of the seating surfaces, the seat ring bore, and the outer contour of the plug is a good practice to prolong service life. For tight shutoff, it is best if there is a difference in hardness between the plug (40 to 46 RC*) and seat ring (56-62 RC). The same can be said for the relation between plug guide and guide bushing. For smaller size plugs, solid ceramic (A_LO_2) or carbides[2] (such as tungsten carbide) have been found to be effective. However, the key here is to streamline the fluid flow in all areas that approach the plug/seat restriction (vena contracta) and to avoid anything that would cause direct fluid impact. Ideally, the valve should only accelerate the fluid.

The actual deceleration with the resultant, very destructive, high turbulence should take place inside a large pipe or vessel downstream—but not close to the valve plug.

As an example of how brief the service life of valve plugs and seat rings can be, I would like to cite Reference 3, in which a 6 in. seat ring made from 316 stainless steel was completely destroyed within a few weeks while handling fine clay in suspension in hot oil at a pressure drop of only 50 psi (I suspect that the valve also underwent cavitation, which, of course, accelerated the destruction), and a fully Stellite® No. 1-coated $1\text{-}\frac{1}{2}$ in. diameter valve plug was completely worn away by uranium ore slurry within two weeks even though the pressure drop was only 100 psi.

Remember, the hardened and expensive valve trim will only delay the destruction, not prevent it. In some cases it has been found to be more cost effective to buy the cheapest valve possible and to replace it on a regular basis. This has been the case with the control of bauxite slurry where a simple carbon steel butterfly valve is used and periodically replaced.

CAVITATION

This process can be avoided in most cases by selecting a high F_L type trim and by perhaps relocating the valve to a point of higher

*Rockwell C-Scale Hardness

static pressure, by placing an additional restrictor downstream or by injecting air or gas into the valve. Selecting a hardened trim will only prolong the service requirements and, of course, will increase the revenue of your vendor from the sale of spare parts (see also Chapter 12).

EROSION BY WET STEAM

On one side of the temperature scale is condensate, which attacks carbon steel. Use only chrome-moly alloys, bronze, or stainless steel as valve body material, or make sure a gasket is placed under the stainless steel seat ring in case a carbon steel body is employed. On the other side is superheated steam, which can discharge water droplets during isentropic throttling in a valve. Here, the use of martensitic, chromium stainless steel is recommended, especially AISI Type 440c, which can be hardened to about 58 RC. For lower temperatures (<750°F), 17-4 pH stainless steel is also useful if heat treated usually at 1075°F, yielding a hardness of about 38 RC. Watch the pH level of the steam. A caustic condition (>7 pH) can cause stress corrosion on parts harder than 35 RC! Incidentally, this also applies to "sour" (sulfur-containing) natural gas.

There is no hard and fast rule for when or under what pressure conditions a trim should be hard faced. Remember, gaseous fluid undergoes sonic velocity in the orifice (typically 1350 feet/second for steam) regardless of whether the pressure drop is from 3000 psi to 500 psi or only from 15 psi gage to atmosphere.

The reasons for trim wear are more subtle. Is the steam handled by the valve fairly wet because the valve is far removed from the boiler? Does the fluid contain a solid catalyst? Does the fluid tend to crystallize under certain conditions? Can two-phase flow (liquid plus gas) exist? Does natural gas coming from a well contain sand? These are all questions that should be asked when specifying the valve material. Otherwise, the general-purpose trim material, i.e., AISI type 316 stainless steel will do just fine with clean liquid or gases even if the pressure drop exceeds 1000 psi, and even though most handbooks limit the allowable ΔP to between 150 and 200 psi.[3]

A final word regarding guide material. For globe valves and low pressure service (< 500 psi ΔP), both plug or stem guide bushing can be 316 stainless steel as long as the 316 stainless steel plug surface or the 316 stainless steel stem surface is somewhat work hardened by suitable machining or roller burnishing techniques. Heavier valve types use type 440c stainless steel guide bushings or, for corrosive service, bushings made of or faced with a cobalt, chromium, tungsten alloy such as Stellite®.

For rotary valves, TFE-dispersed glass fiber bushings work well up to 350°F and TFE-dispersed bronze bushings to 450°F. Other suitable bearing materials are Nitronic 60®, a stainless steel alloy suitable to 800°F or, if properly retained, a solid graphite bushing. For higher temperature (to 1250°F) and/or corrosive service, use a cobalt, chromium, tungsten alloy, or graphite.

To recommend suitable materials for the infinite variety of corrosive chemicals is not possible. However, when in doubt, specify a PTFE (Teflon®) lined valve, which is usually your best bet. Here your choice is between globe style valves with C_v ranges from 0.001 to 100, and rotary valves for larger capacities.

References

1. *Compass Corrosion Guide II*, "A Guide To Chemical Resistance of Metals and Engineering Plastics," La Mesa, CA: Compass Publications.
2. Kennicott, W. L., "Use of Hard Carbide Components in Petrochemical Processing," ASME PAPER 61-SA-51 (1962).
3. Wing, Paul, "Selecting Control Valve Trim for High Pressure Drop and Erosive Service," *ISA Journal*, Vol. 2, pp 332–337, Sept. (1955).

12

CONCERN FOR THE ENVIRONMENT

WILL MY VALVE BE TOO NOISY?

The three basic types of noise sources in valves are: (1) mechanical vibration, (2) cavitation, and (3) aerodynamic noise.

Noise caused by resonant vibration of a valve plug within the radial clearance between plug guide and guide bushing, or between plug skirt and seat ring, is usually the result of severe fluid turbulence and can easily exceed 90 decibels of sound level at a frequency of 3 to 7 kHz. This phenomenon seems not so much related to pressure drop (fluid velocity) itself but rather to unstable fluid flow such as flashing or cavitating water. Fortunately, this is a rather rare occurrence. Based on my experience, resonance occurs more often where the undamped natural frequency of the mass m (weight of valve plug, lbs/32.1) is high in relationship to the stiffness K_n (the perpendicular deflection rate lbs/inch) of the valve stem, i.e., where the undamped natural frequency f_r is below 5000. Now $f_r = 0.16(K_n/m)^{1/2}$ cycles/second. A hollow plug will reduce the mass m, while a shorter unsupported stem length or a larger stem diameter will increase K_n—

all positive remedies. Decreasing the guide clearance will reduce the amplitude but is no sure cure for the problem that can lead to fatigue breakage of stem or plug guide.

The OSHA limit of 8 hours of exposure at 90 dB is measured on the A-weighted scale, which is patterned after the frequency response curve of the human ear; hence, limit = 90 dBA. This is one criterion. On the other hand, noise is nothing but audible vibration, and vibration can cause damage to valves and piping.

Even if you ignore the OSHA limit by putting the valve underground, the pipe will typically break at a sound pressure above 130 dBA[1] measured one meter from the pipe wall. Therefore, tolerate no more than 110 dBA without using "low noise trim" even when the valve is buried!

Don't read on if your valve, handling gas or steam, has an energy coefficient (EC) of less than 1000. This number is derived by multiplying your absolute inlet pressure in psia by the calculated valve C_v and F_L numbers. For example, you reduce 150 psia steam and the required valve $C_v = 125$ at an F_L factor of 0.9. Here your EC = 150 × 125 × 0.9 = 16875—definitely a noisy valve. On the other hand, a valve with a C_v of only 6 has an EC of 6 × 150 × 0.9 = 810 and, therefore, a sound level of less than the OSHA limit of 90 dBA, regardless.

ISA published a reasonably accurate prediction method in ISA-S75.17, Control Valve Aerodynamic Noise Prediction, which was based on modified free jet noise theories.[2]

The 46 equations should not scare you if you possess a full size computer. However, to make life easier, I have condensed the document to a handier level:

Where:

$$L_a = 5 + L_{pi} - T_L + L_G \qquad \text{dBA} \tag{12-1}$$

$$L_{pi} = 25 + \eta_m + 10 \log (C_v F_L P_1 P_2 / D_i^2) + F_G \tag{12-2}$$

η_m is a modified (not the real) acoustical efficiency factor[3] (remember, this is a shortcut), and D_i is the inside diameter of the outlet pipe in inches.

Table 12-1. Important Pressure Ratios (P_1/P_2) for Air

F_L	0.5	0.6	0.7	0.8	0.9	1.0
α	0.60	0.63	0.69	0.75	0.85	1.00
P_1/P_2 critical	1.13	1.20	1.30	1.43	1.61	1.89
P_1/P_2 break	1.95	2.07	2.24	2.40	2.76	3.25
P_1/P_{vc} crit. $= 22\alpha$	13	14	15	16	19	22

P_{vc} crit. $= 0.53\ P_1$

If P_1/P_2 is less than P_1/P_2 critical, i.e., subsonic (see Table 12-1):

In Regime I, i.e., $P_1/P_2 < P_1/P_2$ critical,

$$\eta_m = 26 \log [(P_1 - P_2)/(P_1 F_L^2 - P_1 + P_2)] + 10 \log (F_L^2) - 40 \text{ dB} \quad (12\text{-}3)$$

For Regimes II and III, i.e., $P_1/P_2 > P_1/P_2$ critical, flow is sonic up to Mach 1.4:

$$\eta_m = -40 + 37 \log [(P_1/P_2)/(P_1/P_2 \text{ critical})] + 10 \log (F_L^2) \quad \text{dB} \quad (12\text{-}4)$$

For Regime IV, i.e, $P_1/P_2 > P_1/P_2$ break, jet velocity is above Mach 1.4:

$$\eta_m = -31.2 + 10 \log (F_L^2) + 10 \log [(P_1/P_2)/(P_1/P_2 \text{ break})] \quad \text{dB} \quad (12\text{-}5)$$

For Regime V, i.e., $P_1/P_2 > 22\ \alpha$, jet velocity is above 3 and acoustical efficiency is constant:

$$\eta_m = \eta_m \text{ Max. of Regime IV} \quad (12\text{-}6)$$

NOTE: P_1/P_2 critical $= P_1/(P_1 - 0.47 F_L^2 P_1)$, and F_G is a gas property correction factor, $F_G = 10 \log [(C_0)^4 \times (G_g)^2]$.

For

Air + N_2:	$F_G = 122$	Helium:	$F_G = 123$
Steam:	$F_G = 119$	Hydrogen:	$F_G = 122$
CO_2:	$F_G = 121$	Nat. Gas:	$F_G = 120$

P_1 = inlet pressure, psia; P_2 = outlet pressure, psia
C_0 = speed of sound at atmospheric pressure and 60°F, ft/s
G_g = gas specific gravity (air = 1)

Now the transmission loss is:

$$T_L = T_{Lf_o} + \Delta T_{Lf_p} \tag{12-7}$$

where:

$$T_{Lf_o} = 69.5 + 10 \log (rt_p^2/D_i^3) + 10 \log [(P_2/14.7) + 1] \tag{12-8}$$

D_i = pipe inside diameter, inches
r = distance from the pipe = 39 in. + 1/2 D_i
t_p = pipe wall thickness, inches

and

$$T_{Lf_p}; \text{ If } f_p \leq \text{f}_o, \quad \text{then } \Delta T_{Lf_p} = 20 \log \left(\frac{f_o}{f_p}\right) \tag{12-9}$$

If $f_p > f_o$ and

$$f_p \leq 4f_o, \quad \text{then } \Delta T_{Lf_p} = 13 \log \left(\frac{f_p}{f_o}\right) \tag{12-10}$$

$$\text{If } f_p > 4f_o, \quad \text{then } \Delta T_{Lf_p} = 20 \log \left(\frac{f_p}{4f_o}\right) + 7.8 \tag{12-11}$$

where:

$$f_o = 15700/D_i, \text{ based on air} \quad (12\text{-}12)$$

$$f_p = 226 \times M_j/D_j; \text{ if } P_1/P_2 < 3 \quad (12\text{-}13)$$

$$f_p = 316/D_j(M_j^2 - 1)^{1/2}, \text{ above } P_1/P_2 = 3 \quad (12\text{-}14)$$

$$M_j = \left\{\left(\frac{2}{k-1}\right)\left[\left(\frac{P_1}{P_2\alpha}\right)^{(k-1/k)} - 1\right]\right\}^{0.5} \quad (12\text{-}15)$$

The jet diameter, D_j, is a function of the flow area and the valve style modifier F_d (see Table 12-2); $\alpha = (P_1/P_2 \text{ critical})/1.89$ (see Table 12-1).

If you have a multiple hole, low noise trim, then $F_d = 1/(N_0)^{1/2}$, where N_0 is number of exposed holes.[4]

$$D_j = 0.015F_d(C_v F_L)^{1/2} \text{ in feet} \quad (12\text{-}16)$$

Finally, L_G, the pipe velocity correction factor is:

$$L_G = 16 \log\left[1/1 - \left(\frac{0.023 \times P_1 C_v F_L}{D_i^2 P_2}\right)\right] \quad \text{dB} \quad (12\text{-}17)$$

If this is still too complicated for you, then here is my second method, as simple as A(+)B(+)C.

First establish ''A.'' This is the basic noise level of a single-seated globe valve, an eccentric rotary plug valve, a butterfly valve, or any cage valve with no more than four ports per cage, measured with a schedule 40 pipe, 100 psia inlet pressure, and at an opening equal to about 70% of rated C_v shown in Figure 12-1, where the assumed rated C_v number is equivalent to 10 D^2 (i.e., a 2 in. valve having a C_v of 40). Note that Curve ''A'' by itself will cover probably 50% of all noise evaluations. The higher pressure ratio lines are limited on purpose. Beyond the P_1/P_2 ratios shown, you will see more than 0.3 Mach pipe velocities and the longhand method should be used.

Table 12-2. Typical F_d Values of Control Valves, Full Size Trim

Valves Types	Flow Direction	Max C_v/d^2	F_N	F_d @ % OF RATED FLOW CAPACITY (C_v)					
				10	20	40	60	80	100
Globe, parabolic plug	To Open	13	0.28	0.10	0.15	0.25	0.31	0.39	0.46
Globe, 3 V-port plug	To Open	10	0.44	0.29	0.40	0.42	0.43	0.45	0.48
Globe, 4 V-port cage	Either*	10	0.38	0.25	0.35	0.36	0.37	0.39	0.41
Globe, 6 V-port cage	Either*	10	0.26	0.17	0.23	0.24	0.26	0.28	0.30
Globe, 60 hole drilled cage	Either*	8	0.14	0.40	0.29	0.20	0.17	0.14	0.13
Globe, 120 hole drilled cage	Either*	6.5	0.09	0.29	0.20	0.14	0.12	0.10	0.09
Butterfly, swing-through, to 70°	Either	32	0.34	0.26	0.34	0.42	0.50	0.53	0.57
Butterfly, fluted vane, to 70°	Either	26	0.13	0.08	0.10	0.15	0.20	0.24	0.30
Eccentric rotary plug valve	Either	13	0.26	0.12	0.18	0.22	0.30	0.36	0.42
Segmented V-ball valve	Either	21	0.67	0.60	0.65	0.70	0.75	0.78	0.80

Notes: *Limited $P_1 - P_2$ in flow towards center
d = valve size in inches
F_N = valve specific noise parameter, defined by the author as the F_d at a flow capacity (C_v) equivalent to $6.5d^2$. Depending upon pipe size, a lower F_N means less external noise due to higher pipe wall attenuation.

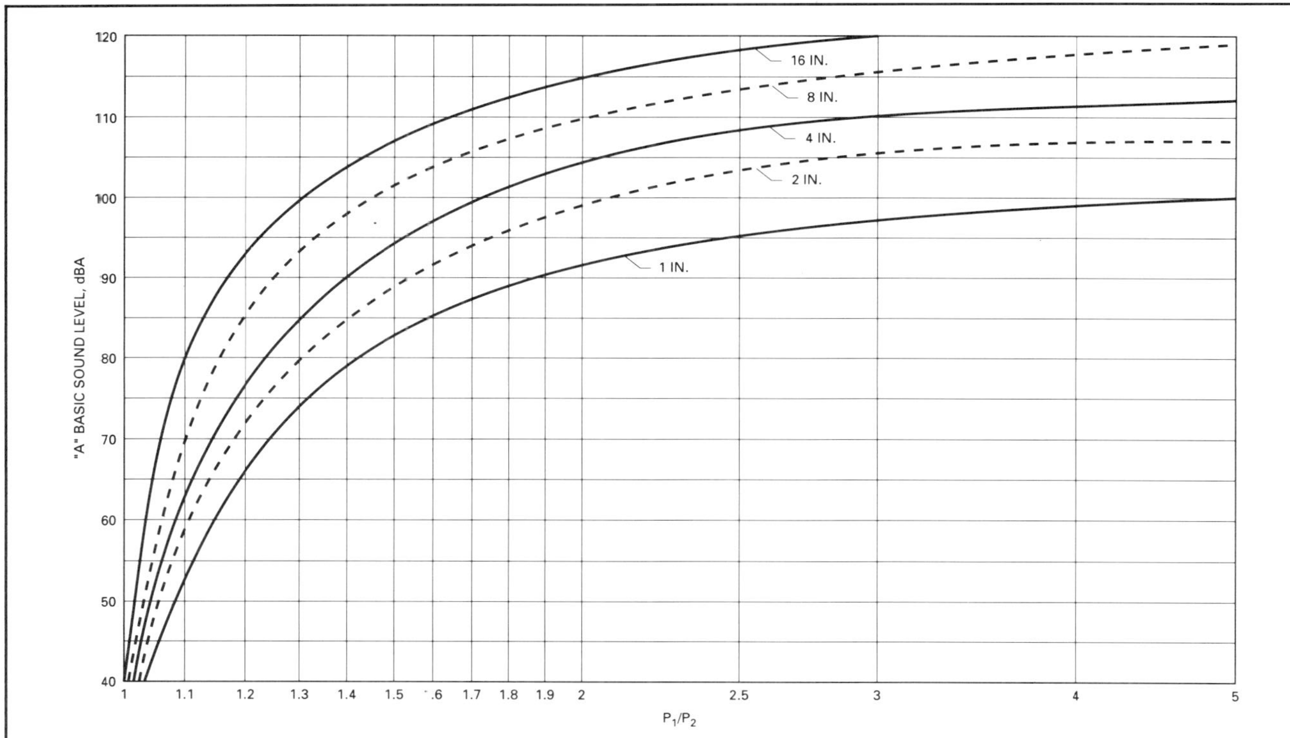

Figure 12-1. Basic aerodynamic sound level in dBA for conventional control valves at approx. 70% of rated flow capacities, at 100 psia inlet pressure, and schedule 40 downstream pipe, measured one meter (3 feet) from the downstream pipe wall.

Table 12-3. Pipe Wall Modifier "C"

Pipe Schedule:	20	40	80	160
C, dBA =	+4	0	−4	−8

Next, we add "B," a correction for inlet pressure if it is different from approximately 100 psia. B = 12 log $(P_1/100)$ dBA. Thus, for 540 psia inlet, B = 12 log (540/100) = 8.8 dBA. Finally, add "C," the correction pipe wall other than schedule 40 (see Table 12-3).

Most gases produce sound levels very close to air. However, you may want to subtract 3 dBA from the total in the case of steam. Using as an example, a 3 in. globe valve with a parabolic plug reducing 540 psia steam to 315 psia in a 3 in. schedule 80 pipe, the C_v calculated is 65 at an F_L number of 0.87. $P_1/P_2 = 540/315 = 1.7$.

From Curve "A" we found no listing for 3 in. but let's read between the lines. At P_1/P_2 of 1.7, we estimate "A" as 97 dBA. Next "B" = 12 log (540/100) = 8.8 dBA. Finally, we add "C" = −4 dBA. The total = 97 + 8.8 − 4 = 101.8 dBA. Subtracting 3 dB for steam gives us about 98.8 dBA at 1 m from the downstream pipe.

Methods of Aerodynamic Valve Noise Reduction

Now to the more common occurrence of "noise pollution," i.e., aerodynamic noise. From the foregoing equations for the estimation of throttling noise levels, it becomes apparent that two basic variables are mainly responsible for the sound level and, therefore, can be varied to achieve a desired noise reduction. They are the jet velocity, M_j, and the internal peak noise frequency, f_p.

The jet velocity is the most important element in the noise generation, yet the most difficult to deal with. Since the velocity is a fraction of the internal orificial pressure ratio P_1/P_o, where P_o is the static pressure inside the orifice, the manipulation of this ratio is the only way to reduce velocity.

In the subsonic Regime I, i.e., $P_1/P_o \leq 1.9$, selection of a valve trim that is less streamlined, i.e., one that has a high F_L number (0.9 to 1.0) will give the lower jet velocity, hence, lower sound level. At higher pressure ratios the only remedy is to use orifices or resistance paths in series so that, ideally, each sub-jet operates in the subsonic regime. Figure 12-2 shows such a valve that has a multi-step resistance trim (the zig-zag path). Assuming an application where natural gas with an inlet pressure of 1500 psia is to be throttled down to 750 psia, and further assuming the valve pressure recovery factor to be 0.9, the internal pressure ratio (P_1/P_o) for a single-stage reduction would have to be $P_1/P_2 = 2.35:1$, yielding a modified acoustical efficiency factor of -34.8 dB from Equation (12-4). However, using a trim from Figure 12-2 with 10 reduction steps, the orificial pressure ratio for each step now is only $1.089:1$ with a modified acoustical efficiency of only -64.5 dB. Ignoring the partial addition of 10 separate sound power sources, this represents a noise reduction of about 30 dB. (The actual noise savings due to added sources and change in frequency, etc., would be about 20 dB).

A less costly way to reduce "audible" throttling noise is to increase the peak frequency by using a multi-orifice trim, as shown in Figure 12-3. Here, the throttling is still single-stage. Only the jets are divided into a number of parallel ports. This lowers the valve style modifier F_d since $F_d = 1/(N_o)^{1/2}$, where N_o is the number of equally sized parallel ports. Since F_d is proportional to the jet diameter D_j, the frequency can be shifted to a higher level.

For example, if $N_o = 16$, as in Figure 12-3, F_d is 0.25, and the resultant peak frequency is $1/0.25$ or four times higher than a single orifice ($F_d = 1$) with the same total flow area. If the peak frequency occurs in the mass-controlled region of the pipe, which is most likely, an overall reduction in the external (audible) sound level of $\Delta T_{Lfp} = 20 \log (f_{p1}/f_{p0.25}) = 12$ dB is achieved. This approach is limited to about $3:1$ pressure ratio. At higher ratios, jets may combine downstream and negate the benefits of high frequency.[5]

A more elaborate trim arrangement is shown in Figure 12-4. The multi-step is combined with the multi-path approach in a layer of disks that have individually cast or etched channels. In this design, we have the beneficial effects of reduction in velocity combined with the benefits of higher transmission loss caused by the increase in frequency. Typical noise reductions are up to 30 dB. Note: Pay attention to the valve outlet diameter. An outlet velocity above 0.2 Mach can negate the beneficial effects of this trim!

Finally, a more economical solution is to couple a static down-

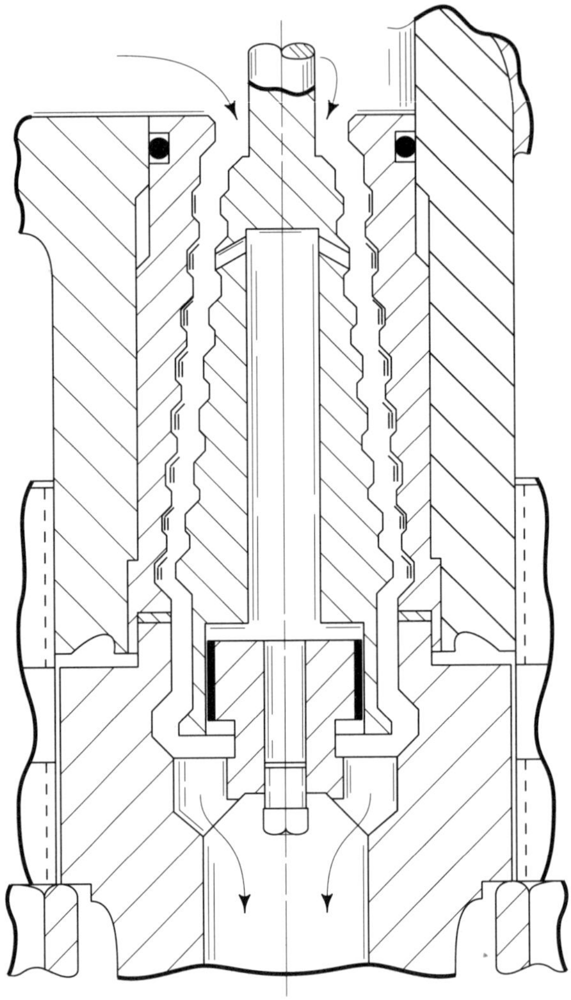

Note: Area enlarges to accommodate change in gas density with lowering of pressure. Benefits are lower throttling velocities.

Figure 12-2. Single flow area, multi-step valve plug. Low noise valve trim using 11 throttling steps with single, annular flow area.

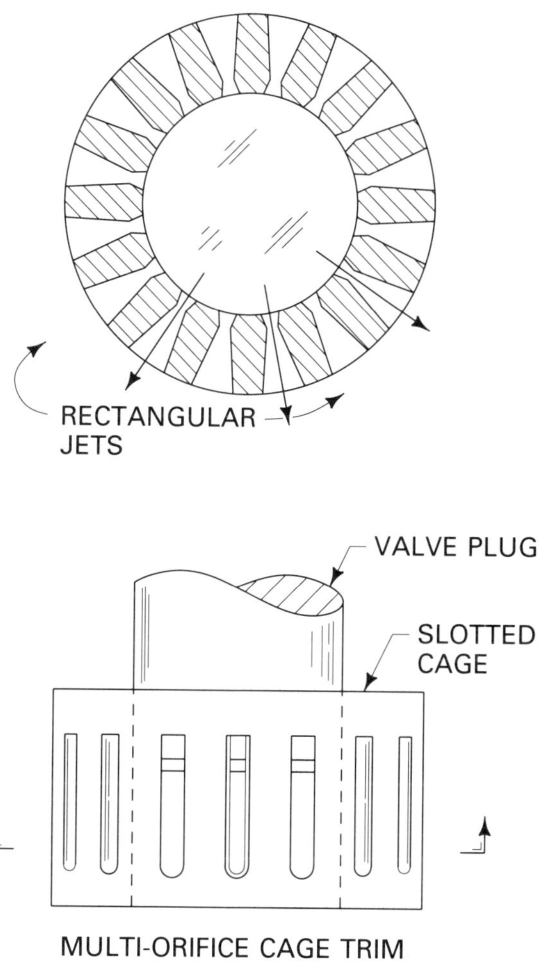

Note: Single-step but multiple flow passages characterize this cage trim. Benefits are higher internal sound frequencies and resultant increase in pipe transmission loss.

Figure 12-3. Multi-orifice cage trim.

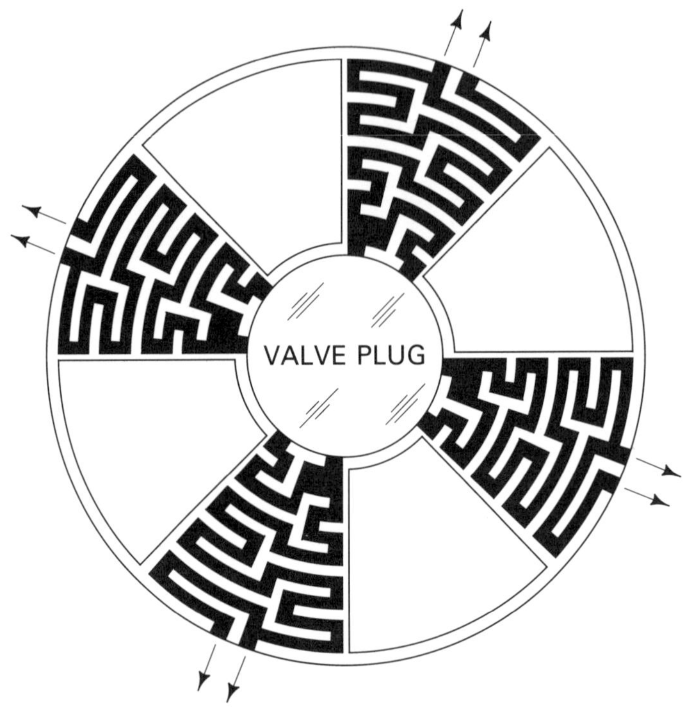

LABYRINTH-TYPE CAGE INSERT

Note: Staked washer-like cage trim featuring a combination of multi-step and multi-path grooved flow channels for reduced velocity and increased frequency benefits.

Figure 12-4. Labyrinth-type cage insert.

stream pressure-reducing device, such as a multiple-hole restrictor, with a throttling valve such as the butterfly valve with a fluted disk shown in Figure 12-5.

Again, the fluted disk will generate multiple, parallel jets with increased noise frequencies similar to the multi-port cage valve. However, most of the pressure energy will be converted by the static plate, resulting in noise reduction at a maximum flow of about 20 dB. At

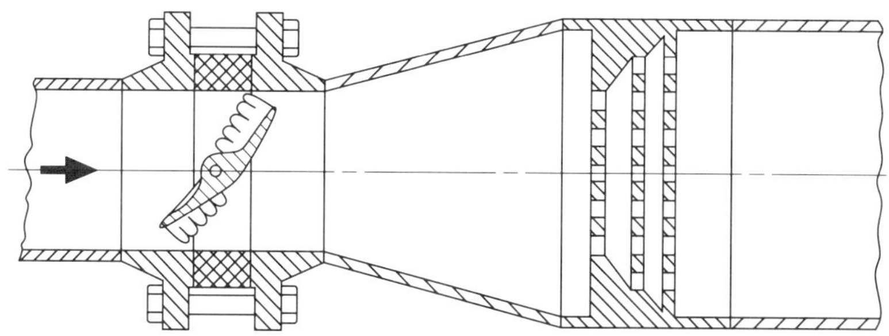

Note: Restrictor is sized to absorb between 90 and 95% of available differential pressure at max. flow conditions.

Figure 12-5. Combination of control valve (fluted, low-torque butterfly valve shown) and multi-stage fluid restrictor.

maximum design flow, the valve is sized to have a low pressure ratio of 1.1 or 1.15 where the rest of the pressure reduction occurs across the static plate, which has inherent low noise throttling due to multi-stage, multi-hole design features.

Placing a silencer downstream of a valve is not cost effective owing to limited noise reduction (approximately 10 dB). This is due to the fact that a good portion of the sound power travels upstream (particularly in so-called "line of sight" valves such as butterfly and ball valves) or radiates through the valve body and actuator. Similar concerns apply to acoustic insulation of the pipe wall where the effectiveness, again, is limited to about 15 dB.

The following is a summary of valve noise abatement procedures:

- If $C_v \times F_L \times P_1$ is less than 1000, then no problem [SL $\leq$ 90 dB(A)].
- 1 in. thick pipe insulation = 5 to 10 dB reduction.
- Doubling pipe wall $\cong$ 6 dB reduction.

- Placing silencer downstream $\cong$ 10 dB reduction.
- With silencer up and downstream $\cong$ 20 dB reduction.
- Multi-port resistance plate downstream of valve = 15 to 20 dB reduction. (Use only 5% to 10% of valve inlet pressure as valve ΔP at maximum flow.)
- Special low noise valve = 15 to 30 dB reduction.

Of course, you can apply a combination of the above methods, which is sometimes more cost effective.

HOW ABOUT CAVITATION?

I used to say, only half in jest, that cavitation is the manufacturer's best friend, since it supplies a steady demand for replacement trim parts. No wonder, since the mechanism of cavitation (violent collapse of vapor bubbles in a liquid following deceleration of the liquid flow and subsequent pressure recovery) can cause impact pressures on the surface of the valve plug, the inner body wall, or the pipe wall up to $1.5 \times 10^3\ N/mm^2$ for water[6] (218,000 psi!). No wonder there is no material known to man that will withstand cavitation damage. A harder material (higher elastic stress limit) will last longer than a softer material. For example, here is the relative life time of some materials (same material loss under the same cavitation conditions):

Material	Relative life
Soft aluminum alloy	1
Type 1020 carbon steel	2.4
Type 304 st. steel	6.5
Hardened 42% CR, 4% Mo steel	13

This means that specifying hardened CR, Mo trim instead of stainless steel may increase the lifetime of your trim by perhaps a factor of two (13/6.5) and is probably not worth the effort.

A wiser course is to avoid (a) pressure conditions that cause cavitation or (b) selecting a valve with a less streamlined trim (high F_L). Unfortunately, process conditions can seldom be altered. However, we can help by choosing a better valve location. For example, placing the valve ahead of a long pipe or heat exchanger will increase the static inlet and outlet pressures. So will placing the valve at ground level instead of in an overhead pipe (static head due to elevation will increase inlet and outlet pressures). All this increases the allowable pressure drop without cavitation problems.

Basically, three flow conditions or pressure ratios in relation to cavitation could concern you:

• *Incipient Cavitation.* This is the pressure ratio at which the first bubbles are formed (detection by acoustic measurement).

$$K_i{}^* = \Delta P/(P_1 - P_v) \text{ or } \Delta P_{\text{cavitation}} = K_i(P_1 - P_v)$$

where:

P_1 = inlet pressure (psia)
P_v = vapor pressure of liquid (psia)

K_i can be as low as 0.10 in some valves, although incipient cavitation can be ignored for all practical purposes (does not produce excessive noise except for very large valves and does not produce sufficient damage).

• *Damage Inducing Cavitation.* This is the important ratio we should consider. Exceeding ΔP damage will result in significant reduction of trim life. By perhaps a lucky coincidence, this ratio most often also defines the onset of 85 dBA noise level with a schedule 40 pipe.[7]

$$\Delta P \text{ damage} = K_D(P_1 - P_v)$$

As an approximation, and for conventional globe and rotary control valves (standard trim):

*Also expressed as σ_i where $\sigma_i = 1/K_i$. In Germany, K_i is called X_{Fz}.

$$K_D = K_S \times K_P \times F_L^3 \tag{12-18}$$

K_S = size coefficient $(1/d)^{0.125}$; d = valve size in inches

Example: An 8 in. valve has a K_S of $(1/8)^{0.125} = 0.77$
K_P is a pressure correction factor, where $K_P = (100/P_1)^{0.125}$
For example, if $P_1 = 225$ psia, $K_P = (100/225)^{0.125} = 0.90$

Assuming further that the 8 in. valve is a butterfly valve, which at an assumed C_v of 640 has an F_L value of 0.75, the coefficient of damage K_D can now be calculated as:

$$K_D = 0.77 \times 0.90 \times 0.75^3$$
$$= 0.29 \text{ from Equation (12-18).}$$

Assuming the fluid is cold water at $P_v = 0.5$ psia, the maximum allowable pressure drop ΔP to avoid noise and cavitation damage is:

$$\Delta P_d = K_D(P_1 - P_v)$$
$$= 0.29(225 - 0.5)$$
$$= 65.6 \text{ psi}$$

Another example: We want to control 180 gpm of a liquid with a vapor pressure at 180°F of 13.5 psia. The inlet pressure $P_1 = 125$ psia and $P_2 = 38$ psia. From our C_v calculation, and assuming $F_L = 0.9 (G_f = 1)$, we find the flow is not critical and C_v required = 19.3. Selecting an $1\frac{1}{2}$ in. globe valve, we find the K_D factor from the above equation to be 0.67.

$$\Delta P \text{ damage} = 0.67(P_1 - P_v)$$
$$= 0.67(125 - 13.5)$$
$$= 74.7 \text{ psi}$$

Now the actual $\Delta P = 87$ psi, so we definitely have a damage and a noise problem that calls for an anti-cavitation trim or placement of a resistance plate downstream of the valve. Let's assume we choose the latter. This increases the outlet pressure at maximum flow to 115 psia, and it decreases the valve ΔP to 10 psi, i.e., less than ΔP damage. The C_v requirement of the valve will increase to 57 and you will have to choose a 2 in. size valve. Now what will happen at half the flow? Since the restrictor has a fixed C_v, the ΔP of the restrictor decreases as $(1/Q)^2$, or by a factor of four. This makes the valve $P_2 = 38 + [(115 - 38)/4] = 57.3$. The valve C_v is now 90 gmp$/(125 - 57.3)^{1/2}$, which is 11 since F_L is still 0.9. The corresponding 2 in. K_D factor is 0.64 and ΔP damage $= 0.64(125 - 13.5) = 71.4$. This is above the limit of the actual $\Delta P = 125 - 57.3 = 67.7$ psi, illustrating that fixed downstream restrictions work if the flow does not vary by more than 2 : 1!

- *Choked Cavitation Flow.* The final cavitation pressure ratio $K_{choked} = F_L^2$. This is the "choked" pressure ratio at which point the flow remains constant with further decrease in P_2, $\Delta P_{choked} = F_L^2(P_1 - P_v)$.

Here the cavitation noise is near its maximum and severe cavitation damage can occur in the downstream piping of rotary valves, for example (also called "super cavitation"). This condition should definitely be avoided except for some short duration emergency conditions.

What To Do About Cavitation

As I mentioned before, try to avoid exceeding ΔP damage. If changing the process conditions is not possible, select an anti-cavitation trim from your vendor's catalog.

Another solution, previously outlined, is the application of a single- or multi-stage resistance plate downstream of your reducing valve. The key to success is proper valve sizing. At maximum flow rate, allocate only 10% of the total pressure drop across the valve and 90% across the plate!

A third solution is to place two identical valves in series. This has the effect of having a two-stage trim. The F_L or K_D of the installation is now $(F_L)^{1/2}$ or $(K_D)^{1/2}$; that is, if a given value F_L were 0.7. The combined F_L is now $(0.7)^{1/2} = 0.84$, thereby giving you a corresponding increase in ΔP capability.

Lastly, cavitation damage and noise can be reduced by gas or air injection, as shown in Figure 12-6. This simple method works as long as P_v is below 7 psia and the fluid can tolerate gas or air entrainment.

A typical application for this method is backpressure control of water from a cooling tower. By placing the point of air injection closely downstream of the point of jet constriction, the differential between the near absolute vacuum (cold water $P_v = 0.5$ psia) and the 14.7 psia atmospheric outlet pressure will force the air through the check valve and into the fluid to act as a cushion for the imploding vapor bubbles. Since the process takes place downstream, there is no effect on the flow capacity of the valve.

Estimating Cavitation Noise

The estimated sound level in dBA is given from Reference 8 as:

$$L_a = 10 \log (C_v F_L) + 8 \log (P_{2\,\mathrm{critical}} - P_2) + 20 \log (P_2 - P_v) + 33$$

where P_2 critical $= P_1 - F_L^2(P_1 - P_v)$.

Note: This is valid only if ΔP is higher than $F_L^2(P_1 - P_v)$ and for schedule 40 pipe.

NOTE: This noise level does not include noise from turbulence or incipient cavitation. The calculated peak noise will occur where the products in the second and the third parentheses are of equal value. Thereafter, the noise will be less, leading to a reversal of noise level. This initial increase in noise with greater pressure drop and then falling when P_2 approaches P_v is actually observed in laboratory tests.

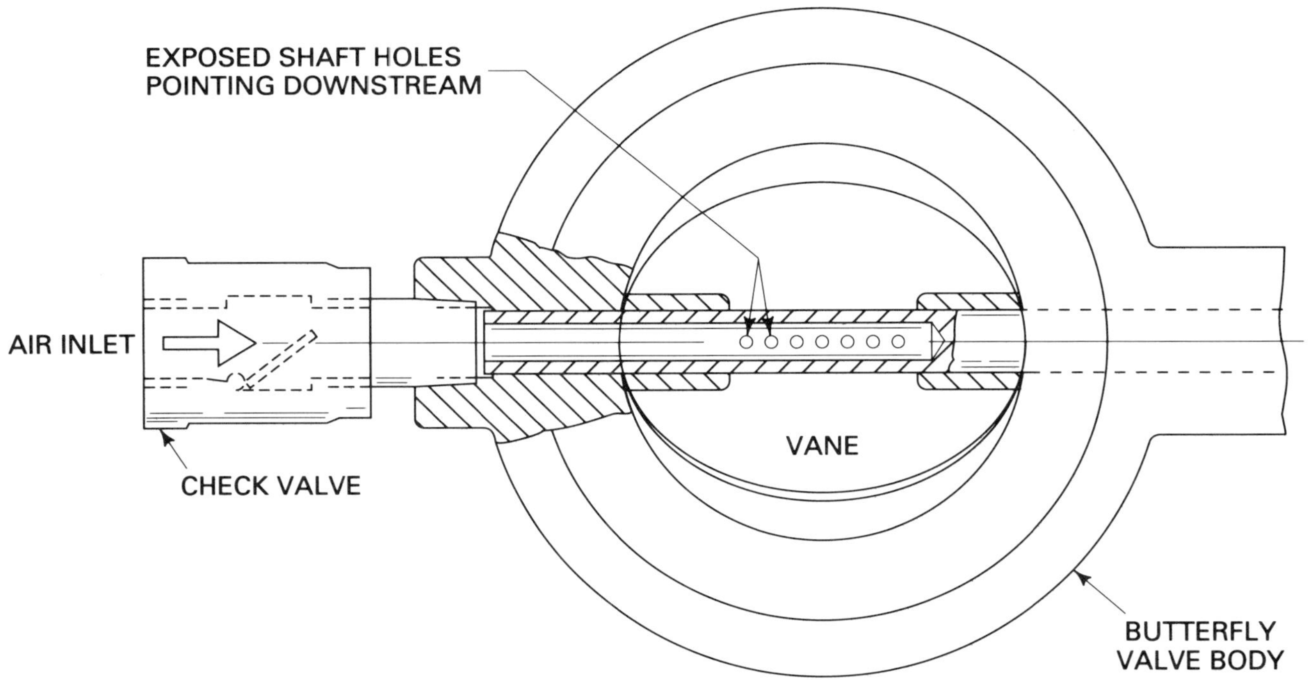

Note: Air enters through the check valve on left due to cavitation-caused vacuum downstream of vane.

Figure 12-6. Air injection-type butterfly valve to reduce cavitation noise and damage. (Courtesy of H. D. Baumann Assoc., Ltd.)

For example: a 6 in. ball valve reducing 100 psia water to 25 psia has an F_L factor of 0.7 at the required C_v of 500.

$$P_v = 0.5 \text{ psia}$$

$$P_{2\text{critical}} = 100 - 0.7^2(100 - 0.5) = 51.2 \text{ psi}$$

$$\begin{aligned} L_a &= 10 \log(500 \times 0.7) + 8 \log(51.2 - 25) \\ &\quad + 20 \log(25 - 0.5) + 33 \\ &= 25.4 + 11.3 + 27.8 + 33 \\ &= 97.5 \text{ dBA} \end{aligned}$$

From Equation (12-18), we know that K_D is only 0.27. This makes ΔP damage = 0.27(100 − 0.5) = 26.9 psi—substantially less than the actual 75 psi. No wonder the noise level is 97.5 dBA. The above example shows that by exceeding ΔP damage, we can incur not only damage but also substantial noise levels!

AVOIDANCE OF LEAKY VALVE STEM PACKINGS

Valve stem packing is one of the most important elements of a control valve, yet one that is not fully understood. One thinks, quite correctly, that a worn stem packing causes external leakage of fluid passing through the valve, but how many of us realize that a tight packing can make the controller go unstable? The combination of low friction and low stem leakage is what makes a good packing. Notice that I said low leakage—not zero leakage—since there is no such thing as zero leakage.

In former times when asbestos was the standard packing material, packing boxes had lubricators whereby the maintenance people could inject grease into the packing, which not only reduced friction but also provided a high viscosity interface between the packing bore and the stem exterior to improve sealability. Any maintenance department had to stock quite a variety of packing greases with varying viscosities (depending on service temperature), specially composed to resist corrosive attacks by certain fluids. Teflon® changed all that and the all-Teflon V-ring—or Chevron packing, preferred by some, and the TFE-lubricated asbestos packing preferred by others—became the industry standard.

While the V-ring packing could not be repaired without taking the whole valve apart (which usually meant shutting down the process), it offered constant stem friction, especially when preloaded by a compression spring. The TFE-asbestos packing or its current successors, the TFE-graphite or TFE-Kevlar® packings, are made in the form of a split-ring and allow the emergency repair of a valve in the line by slipping one or two extra rings under the packing follower. The disadvantage of this type of packing was, and still is, that its tightness depends on the "feel" of the maintenance person, and many a carefully tuned process controller has gone unstable because of added stem friction (increased valve dead time) due to an overtightened stem packing.

Here then are some rules on how to avoid trouble with a valve stem packing:

- Make sure the packing is tight *before* placing the valve in operation. Some manufacturers ship the valve with a loose packing flange or nut.
- If hot fluids are involved, retighten packing after the valve has fully warmed up when first placed in service.
- During installation, do not rotate the actuator (to change positioner location, for example) without loosening the packing prior to moving the actuator.
- Do not allow your control loop or actuator to become unstable for a long time. Remember, even one cycle per second cyclic motion can translate into 2-$\frac{1}{2}$ million cycles of packing wear in the time frame of only one month!
- Do not use packing material that has a high coefficient of thermal expansion when handling steam or hot fluids.
- Total effective packing depth should always exceed the length of valve travel.

- Packing for rotary valves is superior in performance to those for reciprocating stem valves as long as the shaft of the rotary valve is well supported to absorb the perpendicular actuator thrust or the off-center strain caused by a misaligned stem-coupled rotary actuator.
- Misalignment between valve stem and actuator stem causes wear and friction. This can happen when an actuator yoke made from steel or aluminum bends following a drop during shipment. Cast iron yokes cannot do this—they break. Misalignment can also happen when a heavy actuator is mounted in a horizontal position.

While no packing is tight, remember, the permeation rate[9] through *solid* steel alone is within the measurement sensitivity of mass spectrometers. Here are some rough examples of what to expect.

A typical $\frac{5}{16}$ in. stem diameter packing with solid TFE chevron rings (preloaded) exhibited a stem leakage when subjected to 55 psi air at 70°F of about 0.4 scc/min after 100,000 full travel cycles, or about 550 scc/day, which is roughly 2×10^{-5} kg/hr and is still below the 500 parts/million limit. The same valve exhibited a leakage of 1×10^{-5} kg/hr of water under the same conditions. The above test data is based on laboratory measurements and exclude environmental effects such as corrosion or dirt, which will make things worse in real life.

In a stable control loop situation, 100,000 full travel cycles are a great many, and a valve packing may not need to be replaced for a number of years. In other cases, a tight maintenance schedule must be instituted. On the other hand, you may want to consider that the packing wear equivalent to $100{,}000 \times \frac{1}{2}$ in. travel cycles is 625 days of operation with an unstable valve having average stem excursions of ± 0.030 in. at a rate of two cycles per minute (= 1.8 million cycles)! This demonstrates again the need for a stable loop and a stable actuator.

When all else fails, specify a packless valve such as a diaphragm valve (metal or elastomer), a pinch valve (elastomer only), a flexible sleeve valve (TFE, plastic, or elastomer), or a bellows seal valve.

BELLOWS SEALS

The use of bellows is one form of avoiding valve stem leakage and a way to meet EPA-recognized avoidance of hazardous leakage from valves (the others are use of diaphragm and pinch valves).

Bellows seals work well if properly maintained and not overstressed. They should *always* be used with a backup packing and a telltale connection between the bellows seal and packing, leading to a pressure switch or pressure gage to inform or alarm the operator in case of bellows failure. The following are causes of bellows failure.

Overpressurization

This is usually caused by subjecting the valve *with* bellows to hydrostatic line test pressures for which typical bellows are not designed (consult the manufacturer). Other sources of excess pressure are pressure waves such as ''water hammer,'' which is caused by sudden valve opening, or thermal expansion of liquid trapped in the valve and pipe due to ambient temperature changes (a relief valve may be required).

Fatigue

This can be caused by excess valve cycles. Note, most bellows are rated for a minimum of 100,000 full travel cycles, which means mnay years of normal throttling service (two *full* open to closure cycles per day will yield 100,000 cycles in 137 years!). However, if this valve is used in an on-off mode at, say, one cycle every five minutes, 100,000 cycles are reached in about one year.

However, fatigue problems can be more subtle. For example, having the bellows subjected to the pressure fluctuations of a positive displacement pump operating at 100 cycles per minute can *partly* deflect the individual bellows convolutions at a rate of 1 million cycles per week and subsequently break the bellows (here a pulsation damper downstream of the pump is the answer). An unstable, cycling electronic control signal to the actuator can do the same damage by subjecting the stem and bellows to millions of minute cycles.

Mechanical Damage

This is usually caused by improper installation or maintenance. The most common problem is twisting of the bellows when the valve stem is turned. A good valve design prevents the possibility of twisting the bellows during assembly or disassembly of the valve. Remember: Instructions are read only *after* things have gone wrong!

Pressurization of the *exterior* of the bellows increases resistance to working pressure by about 50%. However, this requires the retaining bonnet to be pressure and/or corrosion resistant. Adding balancing pressure to the *inside* of the bellows will increase the pressure rating. However, the fluid has to be inert and, in case of liquids, adequate relief must be provided to allow for venting of displaced volume during valve travel.

References

1. Carucci, V. A., and Mueller, R. T., "Acoustically Induced Piping Vibration in High Capacity Pressure Reducing Systems," ASME Paper 82-WA/PVP-8 (1982).
2. Baumann, H. D., "On the Prediction of Aerodynamically Created Sound Pressure Level of Control Valves," ASME Paper WA/FE-28, (1970).
3. Baumann, H. D., "Coefficients and Factors Relating to the Aerodynamic Sound Level Generated by Throttling Valves," *Noise Control Engineering Journal*, pp 6–11, Jan.–Feb. (1984).
4. Baumann, H. D., "A Method for Predicting Aerodynamic Valve Noise Based on Modified Free Jet Noise Theories," ASME Paper 87-WA/NCA-7, (1987).
5. Reed, C., "Optimizing Valve Jet Size and Spacing Reduces Valve Noise," *Control Engineering*, pp 63–64, Sept. (1976).

6. Dobben, T., "Kavitation in Stellglieder," Frankfurt, Germany: Samson A. G.
7. Baumann, H. D., "A Practical Guide for the Cavitation Prevention of Throttling Butterfly Valves," ISA Paper C.1.84-R777, Research Triangle Park, NC: Instrument Society of America (1984).
8. Baumann, H. D., "Control Valve Noise: Cause and Cure," *Chemical Engineering*, pp 120–126, May 17 (1971).
9. Baumann, H. D., "Should a Control Valve Leak?" *Instrument and Control Systems*, August (1966).

13

TWELVE COMMANDMENTS: WHAT YOU SHALL NOT DO!

You shall not:

1. Install a steam valve at the bottom of a down-header.
2. Install a steam valve at the bottom of a down-header without providing a steam trap near the inlet side of a valve.
3. Locate a condensate return valve more than three feet away from a flash tank or condenser.
4. Buy a cage-trim valve for elevated temperature service without elastic support between the cage and bonnet.
5. Buy an I/P transducer with 3–15 psi output signal when the actuator spring of the valve is also 3–15 psi and there is no positioner.
6. Buy a flangeless, wafer-style control valve without a protective shroud where there is a fire hazard.
7. Start up a control valve without first checking the packing tightness.
8. Hydrostatically test a valve with bellows seal without first checking with the manufacturer.

9. Install a single-seated globe valve with spring-diaphragm actuator, flow-to-close, unless it has a small flow trim or it is used only for on/off service.
10. Use carbon steel valve body material for saturated water (condensate) service.
11. Install a valve in a pipe more than twice the diameter of the valve.
12. Let a controller go unstable for too long; the result may be a leaky stem packing.

14

THE BUS SYSTEM TO THE RESCUE, OR WHAT THE FUTURE MAY BRING

Will the conventional (throttling-type) control valve cease to exist? Not very likely. Remember, the automobile did not completely replace the bicycle. Obviously, it goes against our grain to "waste" energy in a throttling valve. However, despite attempts over the centuries to invent "perpetual motion," i.e., a device that works without "using" energy (converting energy to a higher entropy level), we have found none. The same is true for the control of fluids in a pipe. Basic laws of physics still apply.

Attempts have been made to obtain control more elegantly. For example, by applying electric current to a fluid,[1] you can reduce the flowing quantity by changing the turbulent or laminar flow pattern (see Figure 14-1). This works like a magnetic flowmeter in reverse; however, it is not a very practical approach. Speed-controlled pumps have been tried with limited success. So what is the future?

Well, electronics will bring important changes as far as actuating software and hardware are concerned. The biggest impact will be an internationally standardized fieldbus (hopefully, ISA-S50[2]) that will replace the current analog 4–20 mA signal with a digital communication signal between the controller or computer and the valve positioner. The benefits will be increased functionality (for example, the

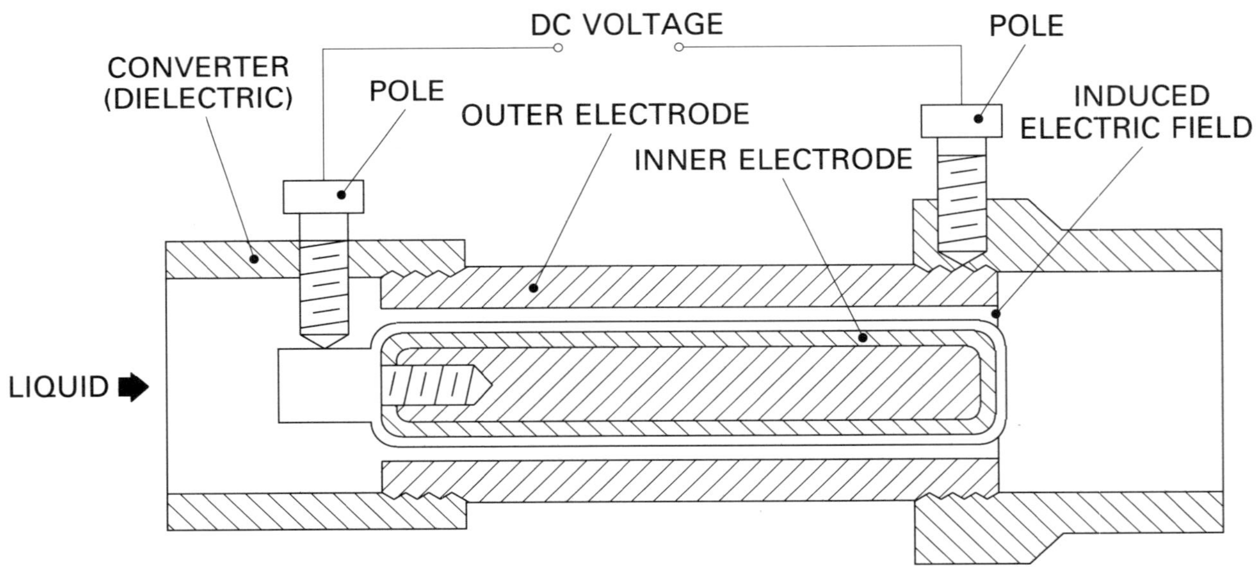

Note: Increase of DC voltage reduces flow velocity and, hence, flow rate.

Figure 14-1. Solid-state control valve.

valve positioner could have a PID function chip and do all the control based on the sensor input via the bus), reduced hardwiring costs, and simplified data transfer with less noise and increased accuracy. At the same time the bus can also transmit important data back to the computer regarding the valve position, for example. Other sensors at the valve can inform about seat leakage, stem packing friction, actuator friction, air leakage in the actuator, and so on. However, I pity the poor operator who must be absolutely overwhelmed by monitoring perhaps ten continuous data input channels per valve in a typical process plant having 500 individual loops! So, selective use of this option is certainly in order. Being informed of a higher than usual valve dead band can explain sudden loop instability and may trigger preventive maintenance, for example.

More serious problems exist with the amount of electric current a bus can handle without losing its "intrinsically safe" label. Of course, a low current means fewer valve positioners on the line. The solution is to multiplex (time share) so that only one positioner in the loop receives a signal change at any given time. This will require a memory device or pneumatic lock-up device to freeze the valve positioner until the next "switch on," which in turn might slow the valve travel speed down to a level that may be unsuitable for flow or pressure control. This is one example of a number of problems that must be addressed prior to the wide acceptance of the bus system. The likely solution would be having separate cables to supply electric power to local actuating devices. This, of course, negates some of the advantages of having a bus system, namely, savings in cable lines to individual components. Nevertheless, there is a basic goal that can be reached, i.e., tie in all local loop components such as sensors, controllers, and valves into an independent local working system (distributed control system) to be connected only for purposes of supervisory control and data logging to a central computer network. This will indeed result in major cost savings not only in cabling but also in the amount of data being transmitted.

"Smart valves," on the other hand, will have a future. The trouble is, what is a smart valve? My preferred definition is what I call "the ultimate in distributed control." Let me explain: First you have to realize that every control valve controls the rate of flow passing through it. Nothing else! The difference between, say, a level control valve and a pressure control valve lies only in the fact that the first one employs a level transmitter instead of a pressure transmitter. If the identical valve handling the same flow is in a loop where the feedback comes from a temperature transmitter, then we have a temperature control valve.

So, once we realize that each valve controls only the rate of flow, then we can go a step further and create a "closed-loop" flow control system with the valve as its centerpiece. Referring to Figure 14-2, a flowmeter, preferably a vortex meter (for rangeability and a wide range of applications, including steam), is attached to the valve inlet (point of lowest turbulence).

A flow controller is either attached to the flowmeter or is made part of the I/P positioner, while the set point for the flow controller is given by the main process controller or computer (shown as part of a temperature control system). The computer sets the flow rate to achieve the ultimate controlled temperature. As long as there is no disturbance in the system, everything will be fine. However, let's assume the heating medium passing through the valve is steam, and, due to demand elsewhere, the inlet pressure drops by 10%. This can cause a sudden reduction in the steam flow rate by 10%, assuming choked flow. In a conventional temperature system it may take several minutes* for the controlled temperature to begin to drop and for the computer to react and open the valve further. This leads to an upset in the controlled temperature.

With the smart valve, on the other hand, the flowmeter will sense the change in flow immediately, will make the valve increase the travel, and will restore the flow *before* the usually slow thermodynamic effects make themselves felt downstream. Thus, the temperature stays at the set point, and the computer probably never even detects the upset in steam pressure. This is, of course, a form of cascade control, except that good portions of the loop are now in a tidy, compact package. Despite the added cost (and perhaps added wiring), a case can be made for the use of such a smart valve in critical loops where the valve can absorb process disturbances before they can affect the ultimate controlled variable.

---

*For example, the temperature measurement lag of a $\frac{1}{2}$ in. bulb thermal well sensing a heavy liquid at low velocities can vary between 60 and 180 seconds!

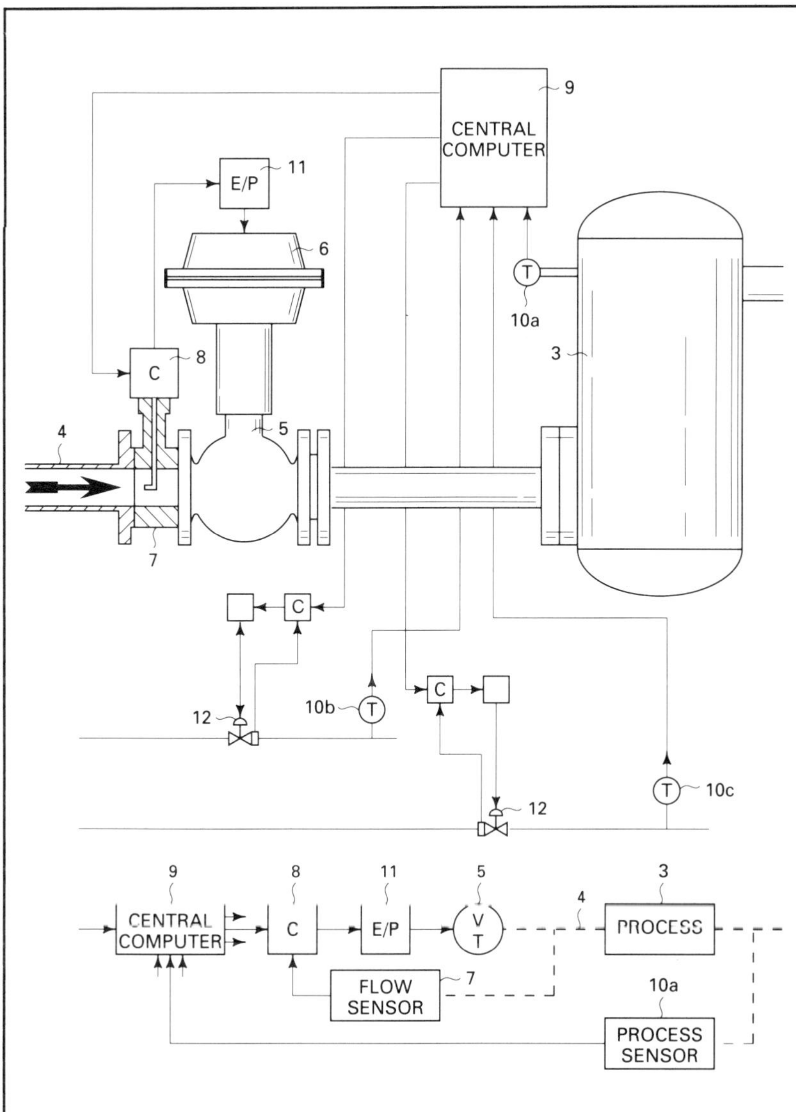

Figure 14-2. Patent drawing (by Author) of distributed control system where the control valve, in combination with a vortex-type flowmeter, fulfills primary flow rate controlling function in a temperature control loop.

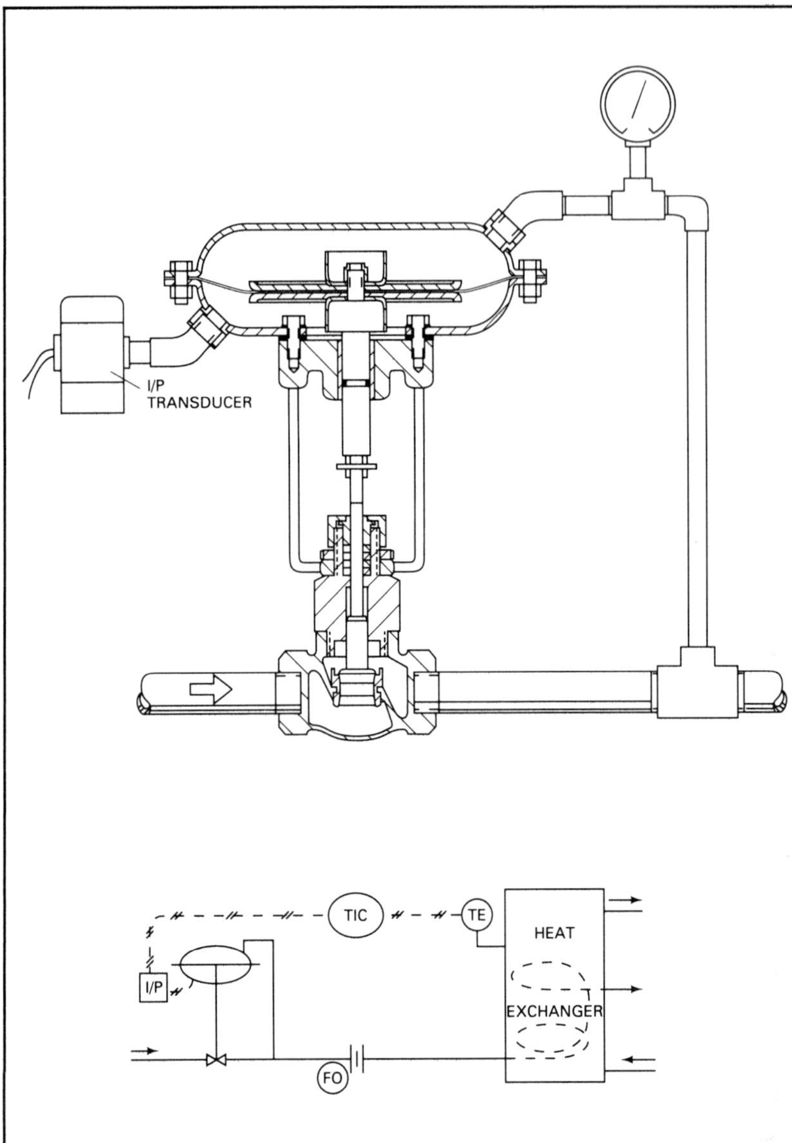

Figure 14-3. Air-loaded regulator serves as a flow controller in conjunction with a fixed or variable orifice (FO) as part of a temperature control loop.

Another favorite of mine, and of a few others,[3] is to resurrect the good old pressure regulator with its excellent frequency response and high flow range. Using a temperature control system similar to that described above, the regulator shown in Figure 14-3 can have a set pressure determined by the air signal from an I/P transducer whose signal comes from a temperature controller (see loop diagram). This regulator controls the pressure of a heat exchange fluid such as oil or steam ahead of an orifice (FO) or an adjustable restrictor. Sized correctly, the combination of orifice area and pressure ahead of the orifice will effectively control the rate of flow (and, therefore, Btu) to the heat exchanger. Any demand for increase in temperature is met by the controller increasing the signal to the I/P transducer, which, in turn, increases the air load below the actuator diaphragm and causes the valve to open, thus increasing the pressure drop across the orifice and thereby increasing the flow of heating fluid. Any local disturbance, such as a sudden decrease in supply pressure, is now sensed by the valve diaphragm, which leads to a nearly instantaneous correction without the disturbance being noticed by the temperature transmitter (TE) at the heat exchanger. We thus, again, have a "smart valve" with nearly the same performance characteristics as those in Figure 14-2 but at a fraction of the cost.

References

1. Baumann, H. D., "Important Trends in Control Valves and Actuators," *Instrument & Control Systems*, pp 21–25, Nov. (1975).
2. Cating, C. E., "Control Valve Industry Update—1990," *Control*, pp 34–38, March (1990).
3. Driskell, L., "Save Money by Combining Regulators, Distributed Control," *Power*, April (1985).

BIBLIOGRAPHY

Beranek, Ver, *Noise and Vibration Control Engineering*, NY: John Wiley & Sons (1991).

Driskell, Les, *Control Valve Selection and Sizing*, Research Triangle Park, NC: Instrument Society of America (1983).

Liptak, Bela A., *Instrument Engineers Handbook*, Rev. Ed., NY: Chilton Book Co. (1985).

Lyons, Jerry L., *Lyons' Valve Designer's Handbook*, NY: Van Nostrand Reinold Company.

Control Valve Handbook, Marshalltown, IA: Fisher Controls International, Inc.

Control Valve Sizing, Bulletin SM-6, Portsmouth, NH: H. D. Baumann Assoc., Ltd.

ISA Handbook of Control Valves, 2nd Edit., Research Triangle Park, NC: Instrument Society of America (1976).

"Flow of Fluids Through Valves, Fittings, and Pipes," Technical Paper No. 410, Chicago, IL: Crane Company.

Masoneilan Handbook for Control Valve Sizing, Avon, MA: Masoneilan-Dresser Co.

Standards and Recommended Practices for Instrumentation and Control, 10th Edit., Research Triangle Park, NC: Instrument Society of America (1991).

Valves, Piping and Pipe Line Handbook, 2nd Edit., Surrey, United Kingdom: The Trade and Technical Press (1986).

APPENDIX A

SAVING ENERGY

As pointed out in Chapter 5, one should use only 5% of the total system pressure or 5 to 10 psi as allowed pressure drop across the control valve at the maximum flow rate. This will save a company a great deal of electrical energy for pumps and compressors. It is contrary to the old rule of using about 50% of the dynamic head pressure in order to achieve a more linear installed flow characteristic (see Chapter 7).

Since the power converted in a control valve is similar to that used in a pump to create the pressure in the first place, it can be calculated as

$$P = 4.2 \times 10^{-4} \times \Delta P \times q \text{ in kW}$$

where:

q = flow in U.S. gpm of liquid
ΔP = pressure drop in psi

For example, in a valve application for 120 gpm oil at G_f of 0.8, the extra pump horsepower needed (a bigger motor is needed) if 50 psi pressure drop is used is:

$$P = 4.2 \times 10^{-4} \times 50 \times 120$$
$$= 2.5 \text{ kW}$$

for a valve size of only $1\frac{1}{2}$ inches ($C_v = 15.2$).

By reducing the valve pressure drop to 5 psi and choosing a different pump, an energy savings of $2.5 - 0.25 = 2.25$ kW per hour or up to 20,000 kW per year can be generated. True, the valve size now has to go up to 2 inches to accommodate the higher C_v (48) based on only 5 psi pressure drop; however, the cost savings of about $2,000 per year more than pays for the cost difference between the two valve sizes. Besides, the line size for 120 gpm flow is 2 inches anyhow (see Table A-1), so the cost of pipe reducers is also saved. Another argument to use line size valves only!

Another benefit besides energy (cost) savings is less valve repair, since there is no trim vibration that could be caused by higher pressure drop and no cavitation to wear out the trim.

In a recent article,* Hans O. Engel pointed out that the three largest chemical producers in Germany use about 120,000 pumps with an average power consumption of 3 kW and cooperating with about 200,000 control valves. If all valves are sized more rationally (i.e., as proposed herein), then about half the power, or 180 MW of electric energy, could be saved. This, in turn, translates to a yearly savings of about 1.5 million MWh. It is fair to assume that the *total* U.S. chemical industry is about three times the size of the above German representation. This means a possible savings of 4 to 5 million MWh per year in the U.S.!

SIZING CONTROL VALVES FOR VISCOUS FLUIDS

The F_R derived from equations stated in Chapter 5 for small flow valves do not agree with values given by Figure 1 in ISA-

---

*Engel, Hans O., "Energieökonomie bei der Anwendung von Regelventilen" rtp, No. 4, 1991, pp 188–194 (Germany).

S75.01, Flow Equations for Sizing Control Valves, and yield more conservative results that may lead to oversizing a globe valve by a factor of four in Reynolds numbers below 100. The current ISA F_R curve is based on a limited number of tests with full sized globe valves while my proposed values are based on air and water tests with small flow valves only.

From data published in Perry's *Chemical Engineers Handbook*, sixth Edition, and Table 5-15, it appears that more streamlined valves (i.e., angle valves and gate valves) exhibit F_R factors close to my curve, while disk and plug-style globe valves (due to their more abrupt flow pattern) tend to favor the current ISA F_R values when corrected with the proper F_d values.

From this data, it appears that for any given Reynolds number, the "pseudo C_v", i.e., $C_v \times F_R$, is identical. This leads me to propose the following relationship for the truly laminar flow regime, i.e., Re_v below 30:

$$F_R = 0.7\, d^2 \sqrt{Re_v}/C_v$$

where C_v is the rated, wide open flow coefficient of the valve, and d is the valve diameter in inches (assuming full size trim). The resultant curves are plotted in Figure A-1 and should be used to correct all control valves for fluid viscosity as outlined in Chapter 5. However, keep in mind that this matter is still under discussion within ISA Standards Committees. As a result, ISA-S75.01 has not been revised.

Table A-1. Flow rates in U.S. GPM Based on Schedule 40 Pipe and 10 Feet/Second Fluid Velocity

Pipe diam.	1″	$1\frac{1}{2}$″	2″	3″	4″	6″	8″	10″	12″
GPM	27	62	100	210	380	900	1600	2300	3500

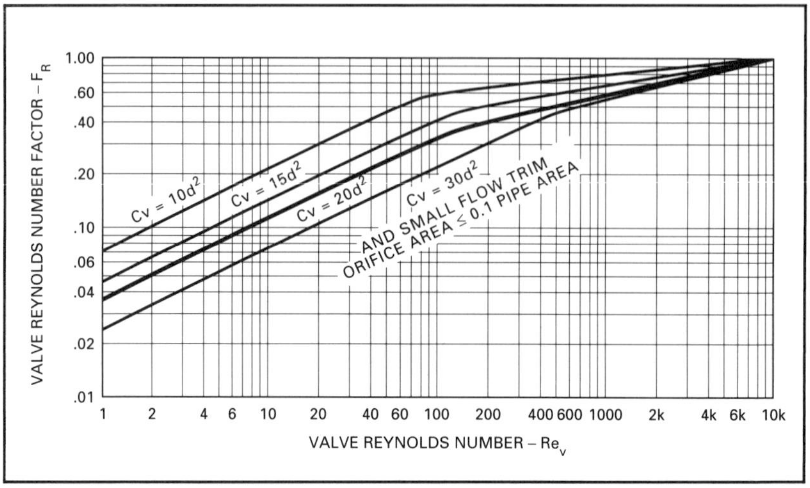

Figure A-1.

Table A-2. Flow Rates in Pounds per Hour of Saturated Steam at a Fluid Velocity of Approximately 150 Feet/Second in a Schedule 40 Pipe

Pressure, psig	Pipe Diameters (Flow in lbs/hr) 1″	1½″	2″	3″	4″	6″	8″	10″	12″
15	200	500	800	1800	3200	5400	12800	20000	24600
30	300	710	1200	2400	4800	9800	19200	30000	36000
60	500	1200	2000	4500	8000	18000	32000	50000	60000
100	750	1800	3000	6800	12000	27200	48000	75000	90000
150	1100	2500	4400	9900	18000	40000	72000	111000	132000
200	1500	3300	6000	13500	24000	54000	96000	150000	180000
300	2200	4700	8800	20000	32000	80000	128000	220000	265000
400	2500	6100	10000	22500	40000	90000	160000	250000	300000
500	3300	7800	13000	30000	52000	120000	210000	330000	400000

APPENDIX B

CONTROL VALVE STANDARDS AND RECOMMENDED PRACTICES

INSTRUMENT SOCIETY OF AMERICA (ISA): 67 Alexander Drive, P.O. Box 12277, Research Triangle Park, NC 27709

ANSI/ISA-S75.01-1985, Flow Equations for Sizing Control Valves

ANSI/ISA-S75.02-1988, Control Valve Capacity Test Procedure

ISA-S75.03-1984, Face-to-Face Dimensions for Flanged Globe-Style Control Valve Bodies (ANSI Classes 125, 150, 250, 300 and 600)

ISA-S75.04-1984, Face-to-Face Dimensions for Flangeless Control Valves (ANSI Classes 150, 300, and 600)

ISA-S75.05-1983, Control Valve Terminology

ISA-RP75.06-1981, Control Valve Manifold Designs

ANSI/ISA-S75.07-1987, Laboratory Measurements of Aerodynamic Noise Generated by Control Valves

ANSI/ISA-S75.08-1985, Installed Face-to-Face Dimensions for Flanged Clamp or Pinch Valves

ANSI/ISA dS75.10-1985, General Requirements for Flanged Clamp or Pinch Valves

ANSI/ISA-S75.11-1985, Inherent Flow Characteristic and Rangeability of Control Valves

ANSI/ISA-S75.12-1987, Face-To-Face Dimensions for Socket Weld-End and Screwed-End Globe-Style Control Valves (ANSI Classes 150 to 2500)

ANSI/ISA-S75.13-1989, Method of Evaluating the Performance of Positioners with Analog Input Signals and Pneumatic Output

ANSI/ISA-S75.14-1984, Face-to-Face Dimensions for Buttweld-End Globe-Style Control Valves (ANSI Class 4500)

ANSI/ISA-S75.16-1986, Face-to-Face Dimensions for Flanged Globe-Style Control Valve Bodies (ANSI Classes 900, 1500, and 2500)

ANSI/ISA S75.17-1989, Control Valve Aerodynamic Noise Prediction

ISA-RP75.18-1989, Control Valve Stability

ANSI/ISA-S75.19-1989, Hydrostatic Testing of Control Valves

ANSI/ISA dS75.20-1989, Face-to-Face Dimensions for Separable Flanged Globe-Style Control Valves (ANSI Classes 150, 300, and 600)

ISA-dRP75.21-1989, Process Data Presentation for Control Valves

ISA-dS75.22-1990, Face-to-Face Centerline Dimensions for Flanged Style Angle Control Bodies

FLUID CONTROLS INSTITUTE (FCI): P.O. Box 9036, 31 South Street, Suite 303, Morristown, NJ 07960

ANSI/FCI 70-2-1976, American National Standard for Control Valve Seat Leakage

FCI 84-1-1985, Metric Definition of the Valve Flow Coefficient C_V

FCI 87-2-1987, Power Signal Standard for Spring-Diaphragm Actuated Control Valves

INTERNATIONAL ELECTROTECHNICAL COMMISSION (IEC), Geneva, Switzerland.
Industrial Process Control Valves:

IEC 534-1, Part 1: Control Valve Terminology and General Considerations

IEC 534-2-1, Part 2: Flow Capacity. Section One: Sizing Equations for Incompressible Fluid Flow under Installed Conditions

IEC 534-2-2, Part 2: Flow Capacity. Section Two: Sizing Equations for Compressible Fluid Flow under Installed Conditions

IEC 534-2-3, Part 2: Flow Capacity. Section Three: Test Procedures

IEC 534-2-4, Part 2: Flow Capacity. Section Four: Inherent Flow Characteristics and Rangeability

IEC 534-3-1, Part 3: Dimensions. Section One: Face-to-Face Dimensions for Flanged, Two-Way, Globe-Type Control Valves

IEC 534-3-2, Part 3: Dimensions. Section Two: Face-to-Face Dimensions for Flangeless Control Valves except Wafer Butterfly Valves

IEC 534-4, Part 4: Inspection and Routine Testing

IEC 534-5, Part 5: Marking

IEC 534-6, Part 6: Mounting Details for Positioner Attachment to Control Valve Actuators

IEC 534-7, Part 7: Control Valve Data Sheet

IEC 534-8-1, Part 8: Noise Considerations. Section One: Laboratory Measurement of Noise Generated by Aerodynamic Flow through Control Valves

IEC 534-8-2, Part 8: Noise Considerations. Section Two: Laboratory Measurement of Noise Generated by Hydrodynamic Flow

IEC 534-8-3, Part 8: Noise Considerations. Section Three: Prediction of Noise Generated by Aerodynamic Flow through Control Valves

IEC 534-8-4, Part 8: Noise Considerations. Section Four: Prediction of Noise Generated by Hydrodynamic Flow through Control Valves

MANUFACTURERS STANDARDIZATION SOCIETY OF THE VALVE AND FITTINGS INDUSTRY (MSS): 127 Park St., N.E., Vienna, VA 22180

SP-25, Standard Marking System for Valves, Fittings, Flanges, and Unions

SP-53, Quality Standard for Steel Castings for Valves, Flanges, and Fittings and Other Piping Components (Magnetic Particle Inspection Method)

SP-54, Quality Standard for Steel Castings for Valves, Flanges, and Fittings and Other Piping Components Radiographic Examination Method

THE AMERICAN SOCIETY OF MECHANICAL ENGINEERS (ASME/ANSI): 345 East 47th Street, New York, NY 10017

B16.5-1988, Pipe Flanges and Flanged Fittings

B16.34-1988, Valves—Flanged, Threaded, and Welding End

B31.1-1986, Power Piping

AMERICAN SOCIETY FOR TESTING AND MATERIALS (ASTM): 1916 Race Street, Philadelphia, PA 19103
Standard Specifications For:

A-105, Forging, Carbon Steel, for Piping Components

A-182, Forged or Rolled Alloy-Steel Pipe Flanges, Forged Fittings, and Valves and Parts for High-Temperature Service

A-193, Alloy-Steel and Stainless Steel Bolting Materials for High-Temperature Service

A-194, Carbon and Alloy Steel Nuts for Bolts for High-Pressure and High-Temperature Service

A-216, Carbon-Steel Castings Suitable for Fusion Welding for High-Temperature Service

A-217, Martensitic Stainless Steel and Alloy Steel Castings for Pressure-Containing Parts Suitable for High-Temperature Service

INDEX